GCSE BIOLOGY

Martin Barker

LONGMAN COURSEWORK GUIDES

Longman

ACKNOWLEDGEMENTS

I am grateful to Geoff Black and Stuart Wall for their encouragement and guidance in editing this book. I am also most grateful for the valuable comments and helpful suggestions made by Alan Jones.

I believe that the clarity and impact of the book have been much enhanced by the illustrations. I would therefore like to thank Kim Symes for her excellent artwork, and John Faulkner for the photographs.

I also acknowledge my gratitude for the considerable assistance in selecting and testing student coursework given by Sean Dempster, of St Christopher School, Letchworth. Thanks also of course to the students themselves for the hard work that has gone into their assignments.

My thanks go also to Dr Nigel Paul for very helpful discussions, and to Catherine Johnston and Henry Barker for their continuing forebearance!

I acknowledge with gratitude permission to print extracts from syllabuses and other documents from the following:
HMSO for the National Criteria (Biology).
London and East Anglian Group,
Midland Examining Group,
Northern Examining Association (including Fig 2.47)
Northern Ireland Schools Examinations Council
Southern Examining Group,
Welsh Joint Education Committee,
International GCSE (including Fig 3.6).
Of course I take full responsibility for the interpretation of the syllabuses and other examining group documents.

Longman Group UK Limited,
Longman House, Burnt Mill, Harlow,
Essex CM20 2JE, England
and Associated Companies throughout the world.

© Longman Group UK Limited 1989
All rights reserved; no part of this publication
may be reproduced, stored in a retrieval system,
or transmitted in any form or by any means, electronic,
mechanical, photocopying, recording, or otherwise,
without the prior written permission of the Publishers.

First published 1989

British Library Cataloguing in Publication Data

Barker, Martin, *1927–*
  Biology – (GCSE coursework guides).
  1. England. Secondary schools. Curriculum subjects: Biology. GCSE examinations.
  I. Title II. Series
  574′.076
  ISBN 0–582–03858–8

Produced by The Pen and Ink Book Company,
Huntingdon, Cambridgeshire

Set in 10/12pt Century Old Style

Printed and bound in Great Britain by
William Clowes Limited, Beccles and London

CONTENTS

ACKNOWLEDGEMENTS

CHAPTER 1	Nature and Importance of Coursework		1
	UNIT 1 INTRODUCTION TO COURSEWORK		1
	1.1 National Standards in GCSE Biology		
	UNIT 2 COURSEWORK IN THE BIOLOGY SYLLABUSES		2
	UNIT 3 HOW BIOLOGY COURSEWORK IS ACTUALLY ASSESSED		3
	3.1 Example of a coursework assessment		
	3.2 How does the assessment take place?		
	3.3 Organisation of assessment sessions		
	UNIT 4 USEFUL READING		7
	UNIT 5 EXAMINING GROUP ADDRESSES		8
CHAPTER 2	Tackling Teacher-Organised Coursework Assessments: Conducting and Interpreting Investigations		9
	UNIT 1 TEACHER–ORGANISED PRACTICAL WORK		9
	UNIT 2 FOLLOWING INSTRUCTIONS		10
	2.1 How are instructions given?		
	2.2 How are marks awarded?		
	2.3 Examples of practical work involving this skill		
	UNIT 3 HANDLING APPARATUS AND MATERIALS		11
	3.1 Handling common apparatus		
	3.2 Handling biological material		
	3.3 Use of the compound microscope		
	3.4 How are marks awarded?		
	3.5 Examples of practical work involving this skill		
	UNIT 4 OBSERVING AND MEASURING		15
	4.1 Observation		
	4.2 Measurement		
	4.3 How are marks awarded?		
	4.4 Examples of practical work involving this skill		
	UNIT 5 RECORDING AND COMMUNICATION		23
	5.1 Accuracy		
	5.2 Presentation of results		
	5.3 How are marks awarded?		
	5.4 Examples of practical work involving this skill		
	UNIT 6 INTERPRETING INFORMATION		34
	6.1 Processing information: calculations		
	6.2 Extracting information		

    6.3   Recognising patterns
    6.4   Making deductions
    6.5   How are marks awarded?
    6.6   Examples of practical work involving this skill

UNIT 7 CARRYING OUT SAFE WORKING PROCEDURES	42
UNIT 8 USEFUL READING	43

CHAPTER 3   Tackling Student-Organised Coursework Assessments: 44
                    Experimental Design and Problem Solving

UNIT 1 STUDENT-DIRECTED PRACTICAL WORK	44
UNIT 2 IDENTIFYING A PROBLEM AND PLANNING AN INVESTIGATION	45

    2.1   Identifying a problem
    2.2   Planning an investigation

UNIT 3 EXPERIMENTAL DESIGN AND THE USE OF CONTROLS	46

    3.1   Experimental design
    3.2   Controls

UNIT 4 FORMULATING AND TESTING A HYPOTHESIS	50

    4.1   Forming a hypothesis
    4.2   Testing a hypothesis

UNIT 5 EVALUATING THE EXPERIMENT AND SUGGESTING IMPROVEMENTS	51

    5.1   How are marks awarded?

UNIT 6 IDEAS FOR INVESTIGATIONS	53

    6.1   Examples of practical work involving this skill

CHAPTER 4   Student Coursework with Examiner Comments   55

UNIT 1 INTRODUCTION	55
UNIT 2 ESSENTIALS OF LIFE – ENZYMES	56
UNIT 3 ESSENTIALS OF LIFE – CELLS	58
UNIT 4 ESSENTIALS OF LIFE – OSMOSIS	60
UNIT 5 DIVERSITY OF LIFE – NATURAL CLASSIFICATION	62
UNIT 6 DIVERSITY OF LIFE – ARTIFICIAL CLASSIFICATION	64
UNIT 7 REPRODUCTION – SEXUAL REPRODUCTION: FLOWERING PLANTS	66
UNIT 8 REPRODUCTION – SEED DISPERSAL	68
UNIT 9 GROWTH AND DEVELOPMENT – GROWTH	70
UNIT 10 GROWTH AND DEVELOPMENT – DEVELOPMENT	73
UNIT 11 HEREDITY AND VARIATION – HUMAN HEREDITY	74
UNIT 12 HEREDITY AND VARIATION – VARIATION	76
UNIT 13 RESPIRATION – INTERNAL RESPIRATION: ANAEROBIC	78
UNIT 14 TRANSPORT SYSTEMS – BLOOD SYSTEMS: THE HEART	80
UNIT 15 SENSITIVITY – ANIMAL SENSITIVITY: THE SKIN	82
UNIT 16 NUTRITION – AUTOTROPHIC NUTRITION: RATES OF PHOTOSYNTHESIS	84
UNIT 17 NUTRITION – HETEROTROPHIC NUTRITION: FOOD ANALYSIS	86
UNIT 18 ENVIRONMENT – ABIOTIC FACTORS: CLIMATIC FACTORS	88
UNIT 19 ENVIRONMENT – BIOTIC FACTORS: POPULATIONS	90

CHAPTER 1: NATURE AND IMPORTANCE OF COURSEWORK

UNIT 1 — INTRODUCTION TO COURSEWORK

Coursework is a very important feature of GCSE; it provides students with the opportunity of having their work assessed during the course. **Coursework** is generally taken to mean any work which is assessed (usually by the subject teacher) as part of the overall examination. However, coursework should ideally occur as a 'natural' part of the learning process, and not just as something 'added on' for the purposes of assessment. In fact, any coursework which forms part of the learning process should be regarded as valid and important, even if it is not assessed!

Assessed coursework is, however, important and it can make a very significant contribution to the overall performance in a subject. For GCSE Biology, coursework is currently allocated 20 per cent of the total marks available by most GCSE Groups, though the NEA allows 30 per cent for coursework.

1.1 National Standards in GCSE Biology

All Biology syllabuses must conform to a set of standards laid down nationally by the Secondary Examinations Council. These standards are referred to as the *Subject Specific Criteria* for Biology. They state that 'at least half of these marks must be awarded on the basis of experimental and observational work in the laboratory or its equivalent.'

The emphasis on 'experimental and observational work' reflects an important aspect of biology as a science. The skills required are basically **practical skills** (see Units 2–4). They cannot adequately be tested in the 'external' examinations at the end of the course.
There are two important implications of this:

▶ Coursework itself is recognised to be an important part of GCSE Biology.
▶ The skills being assessed in coursework may also be tested indirectly by questions in the written papers, too.

Perhaps it is worth remembering at this stage that the theoretical content of biology has in any case largely been derived from the 'experimental and observational work' of many generations of biologists! Anyone studying biology (at any level) is, in a sense, *discovering* important ideas in the subject. Most of these 'discoveries' have already been made, but practical work is a chance to *confirm* previous work as well as to learn important skills. A stated purpose of coursework is indeed to 'stimulate a sense of exploration and discovery'. It is however difficult to undertake experiments and fieldwork in an open-minded way. Students and teachers usually have a fairly good idea of what to expect in any given exercise! This 'problem' is covered more fully in Chapter 3, Unit 2.

ASSESSMENT OBJECTIVES

A student's performance in Biology can only be assessed in terms of what is *measurable*. The Subject Specific Criteria for Biology refer to all 'measurable' aspects of the syllabus as **assessment objectives** (Fig 1.1).

Fig 1.1 Assessment objectives in GCSE

1 Knowledge and understanding	2 Skills and processes	
Candidates should be able to:	Candidates should be able to:	5 analyse, interpret and draw inferences from a variety of forms of information including the results of experiments;
1 demonstrate knowledge and understanding of biological facts and principles, practical techniques and safety precautions;	1 make and record accurate observations;	6 apply biological knowledge and understanding to the solution of problems, including those of a personal, social, economic and technological nature;
2 demonstrate knowledge and understanding of the personal, social, economic and technological applications of biology in modern society;	2 plan and conduct simple experiments to test given hypotheses;	7 select and organise information relevant to particular ideas and communicate this information cogently in a variety of ways;
3 use appropriate terminology in demonstrating this knowledge.	3 formulate hypotheses and design and conduct simple experiments to test them;	
	4 make constructive criticisms of the design of experiments;	8 present biological information coherently.

There are two main types of ability being tested in GCSE Biology; *Knowledge and Understanding* and *Skills and Processes*. Fifty five per cent of *all* the marks in Biology are based on these 'Skills and Processes'.

Coursework Assessments in Biology are based directly on the list of Skills and Processes shown in Fig 1.1. Experimental work and related Observational Skills form *at least* 20% out of this 55%. Even in the 45% of marks available for **knowledge and understanding**, candidates are required to 'demonstrate . . . practical techniques and safety precautions.'

So the Subject Specific Criteria show the importance of skills *and* knowledge as regards the *practical aspects* of GCSE Biology. The 'core content' of all Biology syllabuses must cover four main themes. These are summarised in Fig 1.2. They provide the topic areas in which the various practical skills can be displayed.

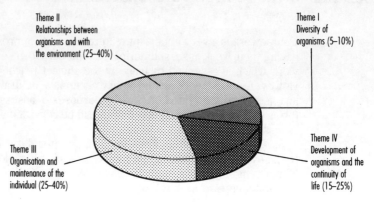

Fig 1.2
Assessment objectives: Themes in the syllabus

- Theme I Diversity of organisms (5–10%)
- Theme II Relationships between organisms and with the environment (25–40%)
- Theme III Organisation and maintenance of the individual (25–40%)
- Theme IV Development of organisms and the continuity of life (15–25%)

UNIT 2 COURSEWORK IN THE BIOLOGY SYLLABUSES

GCSE Biology syllabuses are produced by the six regional Examining Groups. There is also an International GCSE (IGCSE) syllabus for overseas students, organised by the Local Examinations Syndicate at the University of Cambridge. A summary of the coursework requirements for each Biology syllabus is given in Figure 1.3. Since coursework requirements are subject to change – do check the latest syllabus for details.

Fig 1.3
Coursework in the biology syllabuses

Coursework feature	LEAG (A,B,C)	MEG (A, Nuffield)	NEA	NISEC	SEG	WJEC	IGCSE
				EXAMINING GROUPS			
Percentage of total marks allocated ('Weighting')	20	20	30	20	20	20	20
Coursework 'title'	Teacher Assessment (paper 4)	Practical Assessment (component 4)	Centre Assessment of Practical Skills (paper 2)	Coursework (part of 'common component')	Centre Assessment of Practical Skills (paper 2)	Coursework	Practical Assessment Paper 4 (school-based) *or* Paper 5 (practical test) *or* Paper 6 (alternative to practical)
Number of skill areas being tested	3 Categories	6 Skills	6 Domains	5 Skills Areas	7 Skills	5 Skills	6 Skills
Marking	6-point scale (0–5): low, mid, high	10-point scale (0–9)	each of 32 skills worth 1 mark	3-point competence levels (1–3) 0 only awarded 'in extreme cases'	4-point scale (H, J, K, L) worth 0, ⅓, ⅔, full marks for each skill	Skills 1–4 on 3-point scale (1–3): Skill 5 on 3-point scale (2, 4, 6)	10-point scale (0–9)
Frequency of assessment of skill areas	Repeated attempts allowed Categories A & B: at least 2 assessments Category C: at least one assessment	Repeated attempts allowed Skills 1–5: at least 2 assessments Skill 6: one assessment	Repeated attempts allowed: pupil must satisfy *skill criterion* to be given a mark in each case	Repeated attempts allowed: highest level of competence recorded for up to 7 skills in each skill area	15 assessments (equal weighting) Skills 1–4 assessed on at least 3 occasions, Skills 5–6 assessed at least once Skill 7 is given 'overall assessment'	Repeated attempts allowed Skills 1–4 assessed on a minimum of 2 widely-separated occasions. Skill 5 assessed only once No more than 2 skills assessed at a time	Repeated attempts allowed: Skills 1–5; at least 2 assessments Skill 6; one assessment
Completion date for summer exam	30 April	30 April	30 April	31 March	30 April	30 March	30 April

CHAPTER 1 **HOW BIOLOGY COURSEWORK IS ACTUALLY ASSESSED**

UNIT 3 HOW BIOLOGY COURSEWORK IS ACTUALLY ASSESSED

WHEN AND WHERE DOES IT TAKE PLACE?

Practical work has always been an important part of learning in Biology, and the assessment of coursework has given it an added significance. Teachers will often *announce* that a particular practical is to be assessed. The fact that assessment is taking place should not cause you anxiety because:

▶ coursework assessment is not designed to 'catch you out'; it is in fact an opportunity for *strengths* rather than weaknesses to be revealed
▶ you will normally have *several chances* to be assessed in each skill; the best marks achieved are used in the overall assessment
▶ *not all* students are necessarily assessed during a coursework assessment session.

> You can check with your teacher to find out how *your* coursework will be organised

Examining Groups may or may not *require* teachers to inform students that a formal assessment is taking place, though it is generally agreed that this is helpful.

In most cases, the practical work will take place *within normal class time;* many practicals will occupy most of a double period. Some experiments in Biology will need to be set up in one lesson and the observations/measurements made in a subsequent lesson. An experiment to study the factors affecting germination would be an example of this.

Writing-up in class

One of the skills assessed will involve making a *written record* of an experiment (see Ch. 2, Unit 5.2). Teachers will make their own arrangements for this; some will expect all the *writing up* to be done within the classroom or laboratory, perhaps within the same session as the practical work itself. There are several advantages to this:

▶ Observations and measurements may need to be checked; part or all of the experiments may need to be repeated.
▶ Observations and measurements may need to be collected from other members of the group, or class results assembled. Some discussion may need to take place.
▶ Help may be needed from the teacher for instance in the preparation and presentation of results. Sometimes the help given will limit the maximum mark obtainable for a piece of work (Ch.2, Unit 2).
▶ Written work is easily lost or forgotten if not completed fairly promptly. Also, the original purpose and meaning of the experiment may be forgotten.
▶ Teachers can be more confident that written work has been done individually if it is completed under supervision.

It may, however, be appropriate for the written work to be completed at another time or place. This is likely to be so when the processing and interpretation of results is expected to occupy a comparatively long time, such as when several calculations need to be undertaken, or when fairly complex graphs need to be drawn (see Ch.2, Unit 5.2).

Some practical work does not in any case take place within a classroom or laboratory. This applies to fieldwork (see Ch.4 Units 18 and 19) and to project work involving investigations off the school premises. Students may then conduct their own practical work, and the corresponding writing up, in their own time.

WHAT IS ASSESSED?

Coursework assessment involves teachers (in the role as 'teacher-examiners') observing the practical and related written work of *individual* pupils. The teacher gives credit for positive achievement in a limited range of skills in each exercise. The teacher will have decided in advance the skills required for a particular piece of practical work. The pupil will be expected to perform these skills to the best of his or her ability.

Coursework assessment in Biology is based on the range of skills shown in Fig 1.4. These skills are common to all the Examining Groups, though they may be described or emphasised in different ways. Skills 1–3, and also 6, are assessed by *teacher observation*. Skills 4, 5 and 7 involve *pupil communication* (i.e. by written work) which is examined by the teacher. Pupils are required to take considerable responsibility for *problem solving* (Skill 7) and this skill is covered in detail in Chapter 3. The remaining skills are largely based on *practical work* which the teacher organises, and are covered in Chapter 2. An example of a practical assessment exercise organised by the teacher is given below.

CHAPTER 1 NATURE AND IMPORTANCE OF COURSEWORK

Reference in this book	Skill Number	Examining Groups						
		LEAG (A, B, C)	MEG (A, Nuffield)	NEA	NISEC	SEG	WJEC	IGCSE
Chapter 2 Unit 2	1	B: Procedure and use of apparatus	1: Following instructions	3: Assembling common apparatus	2: Following instructions on carrying out procedures.	1: Follow written and diagrammatic instructions	3: Procedural skills	1: Following instructions
Chapter 2 Unit 3	2		2: Handling apparatus and materials	Common apparatus	1: Manipulative skills	2: Handle apparatus and materials	4: Manipulative skills	2: Handling apparatus and materials
Chapter 2 Unit 4	3	A: Making and recording accurate observations	3: Observing and measuring	1: Measurement 2: Observation	3: Observation and measurement	3: Make and convey accurate observations	1: Observational and recording skills	3: Observing and measuring
Chapter 2 Unit 5	4	B: Performing experiments and interpreting the results	4: Recording	4: Recording	4: Recording/ presenting information and interpreting results	4: Record results in an orderly manner	2: Measurement skills	4: Recording
Chapter 2 Unit 6	5		5: Interpreting data	5: Data and its interpretation				5: Interpreting data
Chapter 2 Unit 7	6					7: Carrying out safe working procedures		
Chapter 3	7	C: Designing and evaluating an experiment	6: Experimental design/problem solving	6: Design	5: Experimental design/problem solving	5: Formulate a hypothesis 6: Design experiment to test a hypothesis	5: Formulating an hypothesis and the designing and conducting of an experiment to test it.	6: Experimental design/problem solving

Numbers in the main part of the table refer to skills/skill areas as they appear in the syllabuses

Fig 1.4
Summary of practical skills

3.1 Example of a coursework assessment

A suitable experiment might be to compare vitamin C in different fruit juices (fresh and boiled). This will involve some or all of the following skills; each of these skills is covered in more detail in Chapter 2 in the Units shown.

Skill 1:   Following instructions (Ch. 2: Unit 2)

The teacher will probably describe the purpose of the experiment, then demonstrate the colour change of the dye PIDCP, which detects the presence (or absence) of vitamin C. The teacher may provide a worksheet (Fig 1.5). Pupils can of course seek additional help from the teacher.

Fig 1.5
Sample of worksheet

```
Coursework: Assessment:    23.1.89

Exercise: Experiment to compare vitamin C in
          fresh and boiled fruit juices

MATERIALS NEEDED
 * PIDCP dye
 * range of different fruit juices (4 bottles)
 * boiling tubes (10) and test tube rack
 * measuring cylinder (10 cm³)
 * pasteur pipette
 * bunsen burner
 * test tube holder
 * safety goggles

PROCEDURE

IMPORTANT — read through the instructions first!
Wear safety goggles throughout the experiment

1 Add 5 cm³ of PIDCP dye to a clean, labelled
  boiling tube.

2 Using the pasteur pipette, carefully add one
  of the fruit juices, one drop at a time, to the
  PIDCP dye until you see a colour change.
3 For each fruit juice, record the number of drops
  of juice needed for a blue → clear colour change
  to take place.
4 Repeat steps 1–3 (using fruit juice from the
  bottles). This time, carefully boil each fruit
  juice for 4 mins before testing it. Observe the
  normal safety procedure for boiling liquids.
  Wait until the fruit juice is cool before
  testing it. (Why?). Record the number of drops
  of juice needed, as before.

WRITING UP

1 There is no need to write up the method for this
  experiment.
2 Present your results in a suitable way.
3 Draw conclusions from your results.
4 Suggest any criticisms of this experiment, and
  any ideas for improvements and further work.
```

Skill 2: Handling apparatus and materials (Ch. 2: Unit 3)

This experiment requires a fair amount of organisation. For instance, obtaining and handling the various fruit juices and remembering to rinse out the measuring cylinder after use. If the work area was not properly organised, solutions could be spilt.

Skill 3: Observing and measuring (Ch. 2: Unit 4)

There are several steps involving *measurement* in this experiment; namely the measurement of a 5 cm^3 quantity of PIDCP dye and counting the number of drops of fruit juice added in each case. Each measurement must be done with care, or any conclusions may be unreliable or even invalid. *Observations* may be made as well as measurements, e.g. of the final colour of each solution.

Skill 4: Recording and communication (Ch. 2: Unit 5)

The information (data) obtained by measurements should be preserved in a suitable way. For example, as a table (Fig 1.6). More marks are usually given if *full use* is made of the data; in this example, the difference between the drops used of each fresh or boiled juice is shown.

Fig 1.6
Table of results

Fruit juice	Number of drops of *fresh* juice added to PIDCP dye	Number of drops of *boiled* juice added to PIDCP dye
Pineapple	14	17
Orange	7	10
Grapefruit	9	13
Apple	5	7

Skill 5: Interpreting information (Ch. 2: Unit 6)

The students *interpretation* should be based on the data actually obtained in the experiment. Credit tends to be given for how fully this information is *used*. Specific and/or general reference should be made to *all* the data. In this example, it seems that, in general:
    a) the fruit juices contained *decreasing* amounts of vitamin C in the following order:
        apple → orange → grapefruit → pineapple
    b) in all cases, boiling *reduced* the vitamin C content.

Using the data, the pupil could also make direct comparisons between the fruit juices by using specific references. For instance, the *percentage reduction of vitamin C* could be calculated in each case, and the values compared.

Skill 6: Carrying out safe working procedures (Ch. 2: Unit 7)

Although this skill may be assessed only once or perhaps not at all (depending on the Examining Group), it will be assumed that students will adopt safe practices in *all* practical work, whether it is assessed or not. In this 'fruit juice' practical there are two particular hazards:

▶ Boiling liquids can be dangerous because of the possibility of burns from glass and also because boiling liquid can shoot out of the tube.
▶ Fruit juices might be tempting to drink, but this could be dangerous because glassware may not be clean, and other substances may have been added to the fruit juice.

The nature of an exercise or experiment will determine which skills are assessed. The 'fruit juice' experiment described above provides a good opportunity for Skills 1, 2 and 6 to be assessed. The experiment has also been designed to allow pupils to use their initiative. For instance, boiling the fruit juice (step 4) could be carried out *first*, to allow the solutions to cool (step 1 can be done meanwhile). Skill 3 could also be assessed, though there is not much opportunity to make many *different* observations and measurements; the number of drops of fresh or boiled fruit juice is rather repetitive and unvaried.

The 'fruit juice' experiment provides very limited scope for pupils to demonstrate their ability to *record results*, in the form of drawings, graphs and written results (Skill 4), or to *interpret* them (Skill 5). Skill 7 (*experimental design and problem solving*) is clearly not being used in this exercise at all because it has been organised by the teacher.

3.2 How does the assessment take place?

> **Try to concentrate on the actual coursework itself, not the assessment!**

It is unlikely that *all* students within a class will be assessed during the *same session*, especially if the class is large and if the skills being assessed are mainly practical (i.e. Skills 1–3). If students are working in *groups* (see below), it may be that *one member* from each group will be assessed during any particular session. Also, assessments will normally involve a limited number of practical skills – perhaps two or three skills in each session. Some Examining Groups actually limit the number of skills assessed within any particular session, e.g. to no more than two skills. This makes assessment easier for the teacher, who can then concentrate on a narrow range of skills. It also ensures that pupils are exposed to *several formal assessments*, since there are *several skills* to be assessed during the Biology course.

Most Examining Groups require that each of the skills is assessed *at least twice* (for Skills 6 and 7, it is usually once). There should ideally be a substantial interval between each assessment of the same skill, and the assessment should take place in quite different contexts. This will help to increase a student's chance of demonstrating his or her best achievement.

Examining Groups specify that pupils should be *aware* of the different skills in which they will be assessed, though not necessarily of the particular skills being assessed during each exercise. Examining Groups do not usually instruct teachers to withhold coursework assessment marks from pupils. Individual teachers have their own attitudes to this; in practice, many teachers *do* inform pupils of their marks for *individual exercises* (for instance, by writing marks on written work). However, teachers often prefer *not* to reveal the *total* mark awarded for the coursework component. This is partly because the total mark may not simply be an addition of all the marks awarded for individual assessments.

Another reason for some 'secrecy' is that if pupils *are* aware of their mark, it could lead to complacency or discouragement (depending on the mark!). Of course many pupils may be motivated by *knowing* their mark so practices do vary between teachers and schools in this respect. Coursework is completed by late March or April, just when revision for written examinations should be reaching a critical stage!

Many teachers use a *record sheet* to note down their assessments. For example, the 'fruit juice' experiment described above could be used to assess Skills 1 and 2 on a 10-point scale (this will vary between Examining Groups.) The teacher's record sheet might appear as shown in Fig 1.7. Only a limited number of students have been assessed on this occasion. Other students may have been absent, or perhaps were not yet ready to demonstrate these particular skills.

RECORD SHEET

date: 14.3.89   exercise number: 9   skills assessed: 1, 2
exercise: Fruit juice experiment

Student name	Skill 1	Skill 2	Skill 3	Skill 4	Skill 5	Skill 6
1 R. Bartlett	7	5				
2 N. Boughton	9	8				
3 M. Briant	×	×				
4 I. Brickness	6	3				
5 A. Bullough	7	7				
6 S. Caporn	5	4				
7 T. Crump	3	7				

Fig 1.7
Sample teacher record sheet

Any student being assessed who fails to adequately demonstrate a skill will be given other opportunities. Examining Groups recommend that students who are *not* likely to achieve a particular skill should not be formally assessed for that skill, because the pupil will be unable to demonstrate positive achievement. In such cases, the student will be given a zero mark. Also, failure in one skill should not prevent a student from achieving marks in a related skill. For instance, in the 'fruit juice' experiment, a student who was unable to follow instructions (Skill 1) could observe and measure results (Skill 3) from a teacher demonstration. In other words, skills should ideally be assessed *independently*.

> **Remember that teachers and examiners will be looking for your *best* performance**

The marks for each skill are entered onto individual *record cards* (record sheets) for each student. The best marks are usually transferred to a *summary sheet* for all Biology candidates within each centre (e.g. a school). The actual procedure varies between Examining Groups. The marks available for each skill or sub-skill also vary enormously between Examining Groups. These points are covered in more detail in Chapters 2 and 3.

The *written part* of the coursework assessment (Skills 4, 5 and 7) may be kept separate from other work in *coursework files* for each candidate; these are often retained by the teacher to avoid possible loss. Coursework files may be requested by the Examining Group for *moderation*, i.e. for checking to see that marks are consistent within a school and between schools.

3.3 Organisation of assessment sessions

Coursework assessment, like any other practical work, must be organised according to the *needs* of the students (individually and in groups) and the *availability* of apparatus and other resources:

- All practical groups are likely to consist of a wide range of ability; different students achieve proficiency in different skills at different stages in the course!
- Many practical exercises involve apparatus which is in limited supply, or which requires more than one student to operate it. Students are therefore likely to work in groups; groupwork itself provides opportunities to develop certain skills.

Coursework has to be *organised* by teachers to allow for these complications. This may be done in various ways:

Selected assessment

All students perform an exercise (individually or in groups), with only *certain individuals* being assessed for particular skills.

Circus practicals

> All your coursework practicals will be designed for *your* ability level

All students (or selected groups) perform a series of short-tests or exercises (e.g. for Skills 1–4) arranged as a 'circus'. *Circus practicals* usually consist of numerous short experiments set around the laboratory; pupils progress from one to another, not necessarily in a specific order.

Open-ended practicals

Practical work is often designed in an *open-ended* way, allowing all students to tackle it according to their own ability. This should be true for all skills.

Ability-based practicals

Students are given an opportunity to perform different tasks, according to their ability. More able students may spend more time on project work (Skill 7), whilst those who are less able concentrate on more 'basic' techniques (Skills 1–3).

UNIT 4 USEFUL READING

- *All About GCSE* Maureen Mobley (Heinemann Educational Books, 1986)
- *Biology* Alan Cadogan & Nigel Green (Heinemann, 1986)
- *Biology Passpacks* Alan Cadogan (Longman, 1988)
- *Biology: Revise Guide* Martin Barker (Longman, 1988)
- *GCSE Biology* D G Mackean (Murray, 1986)
- *GCSE: A guide for Parents, Employers and Candidates* Gerry Gorman (Kogan Page, 1988)
- *GCSE: Parent's Guide* Allan Matten (Northcote House, 1988)
- *General Certificate of Secondary Education: A General Introduction* (HMSO, London, 1985)
- *Science GCSE: A Guide for Teachers* (Secondary Examinations Council/Open University, 1986)

*NB: Biology books which cover *practical work* in particular are listed in Ch. 2, Unit 8.

CHAPTER 1 **NATURE AND IMPORTANCE OF COURSEWORK**

UNIT 5 EXAMINING GROUP ADDRESSES

LEAG	**London and East Anglian Group** The Lindens Lexden Road Colchester CO3 3RL (0206–549595)
MEG	**Midland Examining Group** Syndicate Buiding 1 Hills Road Cambridge CB1 2EU (0223–61111)
NEA	**Northern Examining Association** 12 Harter Street Manchester M1 6HL (061–228–0084)
NISEC	**Northern Ireland Schools Examinations Council** 42 Beechill Road Belfast BT8 4RS (0232–704666)
SEG	**Southern Examining Group** Stag Hill House Guildford GU2 5XJ (0483–506506)
WJEC	**Welsh Joint Education Committee** 245 Western Avenue Cardiff CF5 2YX (0222–561231)
IGCSE	**International GCSE** Syndicate Buildings 1 Hills Road Cambridge CB1 2EU (0223–61111)

The names and addresses of the main Exam Groups are listed above. Should you wish, you can write and request an order form to purchase your own copy of the syllabus. You will then have to complete the order form and enclose the cost of the syllabus and postage.

Do check the current coursework requirements as they are subject to change (e.g. MEG for 1990 and NISEC for 1991). Remember to check the *appropriate* syllabus, i.e. the one *dated with the year the examination will be taken*.

CHAPTER 1

TACKLING TEACHER-ORGANISED COURSEWORK ASSESSMENTS: CONDUCTING AND INTERPRETING INVESTIGATIONS

UNIT 1 TEACHER–ORGANISED PRACTICAL WORK

The skills required for practical work organised by the *teacher* are the subject of this chapter. Skills needed for practical work organised by the *student* are covered in Chapter 3. In both chapters, the actual skills specified by the Examining Groups are given, together with advice on how to develop the skills successfully.

Much of the practical work in Biology is planned and organised by the subject teacher. Biology teachers often organise their classwork to involve *both* theoretical and practical work on the same topic. The intention is that each approach, theoretical and practical, is *complementary* to the other. Some Biology courses (for example, LEAG Syllabus C and MEG Syllabus B) are actually based primarily on an experimental approach.

All Biology syllabuses allocate less than half (i.e. 20 or 30 per cent) of the total assessment to coursework. The remaining proportion (80 or 70 per cent) is assessed through written examinations. However, the amount of *teaching time* allocated to practical and theoretical work may be approximately equal. In other words, practical exercises occupy more time than might be expected from the mark distribution. Some practical work is clearly more time-consuming than others, and this may not necessarily be reflected in the mark allocation. Also, some topics in Biology lend themselves more readily to practical work than others. This means that the time spent on practical work may be spread *unevenly* throughout the course.

The *teacher* will decide when a particular piece of practical work is undertaken, for instance immediately before or after a corresponding piece of theory. Teachers will normally explain the purpose of any particular practical exercise. They will often organise the apparatus that will be needed, decide which methods are to be used and the way in which observations will be recorded. This is *teacher-centred practical work* in the sense that the teacher is responsible for planning and organising it. Some practical work may even be undertaken by the teacher, too. Such *demonstrations* can be used to help students to develop new skills.

> **Assessments are made when you are ready to *demonstrate* the skills concerned**

Students need particular *skills* in order to complete practical exercises satisfactorily. GCSE courses allow students to develop and, ultimately, be assessed in these skills. Assessment is normally undertaken when students can demonstrate their *maximum ability* in each of the skills. *Assessments* therefore take place *towards the end* of the course. However *practical work* is useful *throughout* the course because:

▶ it enables students to *develop* the necessary skills, which are eventually required for the coursework assessment
▶ it can promote a deeper understanding of *theoretical* aspects of the subject
▶ experimental aspects of Biology may be the subject of questions in the *written examination.*

There are some 'classic' experiments and investigations in Biology which are quite popular with Biology teachers (and Biology examiners!). Examples of many standard experiments appear throughout this book in Chapters 2–4, and also in the Longman GCSE Revise Guide in Biology. Such standard experiments tend to be used in teaching because they are relatively easy to perform and because they are fairly reliable. They often demonstrate a particular idea in Biology very effectively. However, it should be remembered that there are *many* practical approaches to any given topic in Biology.

All GCSE Biology syllabuses require students to be able to design and evaluate their own experiments. This more *Student-centred* practical work is covered in the next chapter. However, it is fair to say that students learn most of the techniques of experimental design through doing practical work already planned by the teacher. The main skills required by all Examining Groups are summarised in Fig 1.4. These form the basis of this chapter and Chapter 3. The skills are also referred to in relation to examples of coursework in Chapter 4.

CHAPTER 2 TEACHER-ORGANISED COURSEWORK ASSESSMENTS

UNIT 2 — FOLLOWING INSTRUCTIONS

Following instructions is a skill required by all Examining Groups and all (except NEA) give direct credit for it (Fig 2.1). Even if the skill is *not* actually being assessed, it is vital to all teacher-centred practical work.

If a student is unable to follow instructions satisfactorily then help will be given by the teacher; this will be marked accordingly (see Fig. 2.3 opposite). However, a student who receives help in following instructions can at least continue with the rest of the exercise, which may involve the assessment of other skills; for example, recording and interpreting results.

Fig 2.1 'Following instructions': skill summary

LEAG	MEG	NEA	SEG	WJEC	IGCSE	NISEC
Follow the instructions provided.	Follow oral, written and diagrammatic instructions in order to carry out a practical investigation.	Assemble the apparatus from a given set of instructions or diagram so that the apparatus works. Attention must be paid to safety.	Follow written and diagrammatic instructions.	Follow instructions given in written or diagrammatic form, supplemented where necessary with oral instructions.	Follow oral, written or diagrammatic instructions in order to carry out a practical investigation.	A: assemble apparatus (following instructions) B: carry out procedure(s) or follow/complete a sequence of activities.

2.1 How are instructions given?

Instructions can be given by the teacher in various ways:

▶ **written instructions** (see the sample worksheet in Fig 1.5)
▶ **diagrammatic instructions,** including flow charts (Fig 2.2)
▶ **spoken** (oral, verbal) **instructions**
▶ **teacher demonstrations** of practical techniques.

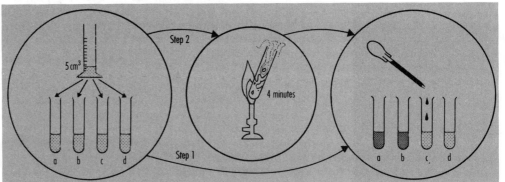

Fig 2.2 Diagrammatic instructions: example of a flow chart

These different forms of instructions are used individually or in combination, according to the nature of the practical work. Teachers are expected to give *clear instructions*, and this is particularly important when the task is complex or potentially hazardous. Students are given credit for working safely; some Examining Groups even assess 'safe working procedures' as a separate skill (Unit 7), though pupils are in any case expected to work safely at all times in the laboratory.

2.2 How are marks awarded?

Marks are awarded in this skill according to how easily the student is able to follow instructions. In practice, this can be assessed in terms of *how much help* is needed from the teacher. The difference in marks for this skill is shown in Fig 2.3.

Teachers will not deliberately make instructions difficult to follow. However, students should concentrate carefully; oral instructions and demonstrations may not be repeated. To be awarded marks in this skill, students should be able to assemble apparatus, perform the experiment and work safely throughout. Any student who does *not* feel able to do this should seek help from the teacher (rather than from another student). GCSE students are encouraged work cooperatively, but for assessment purposes they should if possible be able to work independently as well. Also, other students working in the same group may not themselves have fully understood the instructions.

During assessments, teachers can establish how well *individual* students have understood the instructions by observing them and by asking them questions about what they are doing. Marks are not necessarily deducted if a student simply asks for clarification of the instructions.

Standard	LEAG	MEG	NEA	SEG	WJEC	IGCSE	NISEC
High	Carry out procedure for experiment without assistance	Follow instructions accurately and fluently, taking sensible initiatives		Follow instructions without assistance	Competent in carrying out procedures in accordance with instructions without prompting	Follow instructions accurately and fluently, taking sensible initiatives	Follow a wide variety of instructions, without assistance in a confident and business-like manner
Mid	Carry out experiment with some guidance	Follow routine instructions, with occasional help		Follow instructions with some assistance	Able to carry out procedures with some prompting	Follow routine instructions, with occasional help	Follow flow diagrams and written instructions with minimal assistance
Low	Carry out experiment with constant supervision	Follow experimental instructions given considerable guidance		Follow instructions with considerable assistance	Task only partially completed unless given continuous prompting	Follow experimental instructions given considerable assistance	Follow flow diagrams (accompanied by demonstration) under supervision

Fig 2.3 'Following instructions': skill differentiation

2.3 Examples of practical work involving this skill

The examples given here have all been suggested by different Examining Groups. Teachers may adopt any of these experiments for the assessment of this or other skills. Such experiments are likely to involve several separate steps, which often need to be carried out in a particular sequence. The apparatus used in these exercises are likely 'to involve several components' (NEA); these should be assembled correctly and safely.

Examples include:

- Use of potometer to demonstrate water uptake by a leafy shoot.
- Setting up and using sterile agar plates to demonstrate growth of bacteria.
- Setting up test tubes containing different solutions, and adding to them cylinders of potato to demonstrate osmotic effects (Ch. 4, Unit 4).
- Setting up Visking (cellulose) tubing containing a starch/amylase mixture as a model gut, to demonstrate dialysis/osmosis.
- Setting up different tubes containing a leaf and bicarbonate indicator, to investigate gas exchange during photosynthesis.
- Preparation of a microscope slide with stained onion epidermis or cheek cells (Ch. 4, Unit 3).
- Testing a leaf for starch, as evidence for photosynthesis.
- Comparison of carbon dioxide in inhaled and exhaled air, using a human subject or a small mammal.
- Dissection of a flower, e.g. wallflower (Ch. 4, Unit 7).
- Investigation of the distribution of stomata on leaves, using an imprint technique.

UNIT 3 HANDLING APPARATUS AND MATERIALS

Whilst 'following instructions' requires skills in understanding, *handling apparatus and materials* requires *manipulative skills*. Students must be able to handle apparatus and materials 'appropriately and with care and precision' (NEA). This skill should be demonstrated 'efficiently, effectively and with due regard to safety' (IGCSE) if maximum marks are to be gained. Students are expected to work in an 'organised way' (LEAG), with or without supervision and assistance; students working without additional help will be

Fig 2.4 'Handling apparatus and materials': skill summary

LEAG	MEG	NEA	SEG	WJEC	IGCSE	NISEC
Handling apparatus/setting up and using apparatus	Use various pieces of laboratory apparatus, and biological material efficiently, effectively and safely	Handle apparatus appropriately and with **care and precision** (see Fig 2.13)	Handle apparatus and materials	Handle, assemble and/or use apparatus and/or materials correctly and safely	Use various pieces of laboratory apparatus, and biological materials efficiently, effectively and with a due regard to safety	Handle living organisms with due respect. Use appropriate sampling and collecting techniques. Use a microscope. Use a range of biological tools and general scientific apparatus.

awarded more marks than those who do not. SEG candidates are assessed for this skill in terms of their ability to avoid making minor or major errors.

3.1 Handling common apparatus

Some pieces of apparatus are used in a fairly wide range of practical work in Biology. These include the bunsen burner, the mortar and pestle, and various laboratory glassware including the filter funnel. Simple 'biological tools' are frequently used. They include 'dissecting instruments' such as scissors, scalpels, forceps and also the hand lens.

'General' techniques using the *bunsen burner* include heating solids (for instance, on a mounted needle) or liquids (e.g. in a test tube using a test tube holder). The use of a 'water bath', kept at a fairly constant temperature, is a useful skill for instance in enzyme experiments (Ch. 4, Unit 2). Students should be able to maintain a beaker containing about 200 cm^3 of water within the range 38–42°C (i.e. mean temperature of 40°C) by moving a bunsen burner towards or away from the beaker. This exercise will obviously involve another skill, i.e. measurement. Students may also have an opportunity to use automated water baths (Fig 2.5). It is good practice to *check* the temperature of such baths at intervals, using a thermometer.

> **Try to learn how to use all common apparatus correctly and confidently**

Fig 2.5
Using a series of automated waterbaths

3.2 Handling biological material

The study of Biology, by definition, involves the use of 'biological material'. This may include preserved specimens (often contained in a fluid preservative) (Fig 2.6), parts of skeletons (for instance, skulls and limb bones) and living organisms, or parts of organisms. Examples of organisms might include microbes (e.g. bacteria, yeast cells), invertebrate animals (e.g. woodlice, maggots, earthworms), small vertebrate animals (e.g. mice) and plants (e.g. algae, small flowering plants). Parts of organisms used in Biology teaching typically include fresh material such as flowers, leaves and twigs from various plants, blood and saliva from humans. (*Note: some Local Education Authorities have restricted the use of human body fluids in Biology teaching, due to the risks of HIV ('AIDS') virus transmission.*)

Biological material is used for observation and measurement (see Unit 4), including physiological experiments, e.g. respiration (Ch. 4, Unit 13) and photosynthesis (Ch. 4, Unit 16). Living organisms can also be used for experiments in the growth in individuals (Ch. 4, Unit 9) and variation in populations (Ch. 4, Unit 12).

Fig 2.6
Example of preserved specimens: arthropods

The emphasis in ecological studies (Units 4.19–21) is on the interaction of organisms with each other and with their non-living environment. Observations of organisms in their habitat should not involve too much interference with them. You should try to avoid disturbing plants and animals during ecological studies. This is particularly important for species which are rare or vulnerable.

Measurements of organisms may involve the collection of *samples* (Fig 2.7), for instance to estimate the abundance and distribution of species within and between habitats. Animals should be sampled using 'appropriate sampling and collecting techniques' (NISEC). After observations and measurements have been completed, animals should, if possible, be returned to their original habitat.

Fig 2.7
Sampling animals in a stream

The handling of animals must be conducted with care and respect. *Biologists have a particular responsibility for the living animals which they are studying.* The idea of respect for life derives from an understanding of living organisms, and forms part of the Subject Specific Criteria for Biology. Though it is difficult for teachers to assess 'respect for all life', students of Biology should be able to 'handle living organisms with due respect in both laboratory and field' (NISEC).

CHAPTER 2 HANDLING APPARATUS AND MATERIALS

> You will not be compelled to undertake dissections of animal material against your will

The Subject Specific Criteria for Biology state that GCSE candidates should not be *required* to undertake dissections of animals (or parts of animals). If students prefer *not* to take part in (or simply observe) an animal dissection, they should inform the teacher in advance. Teachers will normally be understanding and may offer an alternative practical exercise for such students.

Students have the opportunity of manipulating tissue using 'biological tools'. This skill may be gained using animal tissue (e.g. sheep's eye and heart dissections) or plant tissue (e.g. a half-flower dissection – Ch. 4, Unit 7).

3.3 Use of the compound microscope

The compound microscope (Fig 2.8) is a specialised piece of apparatus used for examining biological material that is too small (i.e. less than about 5mm) to be seen easily with a hand lens or the unaided eye. Most GCSE Examining Groups expect candidates in Biology to have had an opportunity to develop two related skills involving the microscope:

- preparation of specimens for microscopic examination
- use of a microscope; this may include the drawing of specimens (see Unit 5).

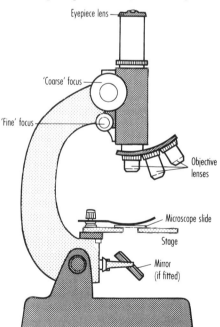

Fig 2.8
A typical compound microscope

PREPARATION OF SPECIMENS FOR MICROSCOPIC EXAMINATION

Specimens normally need to be thin enough to allow light to pass through them (there are exceptions to this, for instance the examination of very small structures in outline, under low magnification). Specimens are mounted on glass microscope slides. In GCSE, students are expected to prepare temporary (rather than permanent) mounts. There are two common methods for preparing *temporary mounts*:

Wet mount

The specimen is placed in a drop of liquid (usually water or a dilute stain); a small square of glass, called a coverslip, is then lowered on top (Fig 2.9). To avoid trapping air bubbles, the best method is to lower one edge of the coverslip whilst holding the opposite edge. The completed mount should be free of bubbles, with no excess liquid on the slide. The specimen should be flat and be a suitable size; a common mistake is to use too much material, so that some of it is not covered by the coverslip. The correct method is used in preparing a slide of onion epidermis in Chapter 4, Unit 3.

If the specimen has not already been stained, a drop of stain can be added under one edge of the coverslip. This is drawn across the specimen by applying a thin strip of filter paper at the opposite edge. This is called the *irrigation method* of staining (Fig 2.10). Iodine (potassium iodide) is a suitable stain for use with onion cells.

Fig 2.9
Wet-mounted specimen: lowering the coverslip

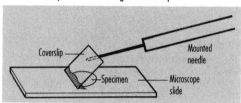

Fig 2.10
Wet-mounted specimen: irrigation staining

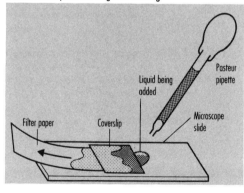

Smear mount

The specimen, a liquid or suspension, is thinly spread across the microscope slide (Fig 2.11) and allowed to dry. This method is appropriate for blood, because individual cells are too numerous to examine if they are not spread out. Red blood cells are clearly visible using this method. White cells are not easy to see because there will be relatively few of them and because they are colourless; a drop of stain (e.g. Leishmann's) can be spread across the slide, as before.

CHAPTER 2 TEACHER-ORGANISED COURSEWORK ASSESSMENTS

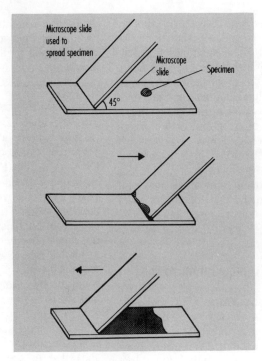

Fig 2.11
Smear-mounted specimen

USE OF THE MICROSCOPE

Microscopes are precision instruments and must be treated with care. Using them is a useful skill in Biology, but this does require practice. There are many different types of microscope currently in use (Fig 2.8 shows a typical model). Students should ask their teachers to demonstrate the operation of the microscopes which are in use in their school or college. There are, however, some general principles in using any standard microscope:

▸ the light system (mirror or lamp) should be set up before attempting to view a slide. The effect of changing the mirror position can be followed by looking down the viewing tube with the objective lens temporarily removed. CARE must be taken not to allow bright sunlight to enter the lens system via a mirror.

▸ The underside of a wet-mounted slide (see above) must be dry before the slide is put on the microscope stage

▸ The specimen must first be 'centralised' within the field of view. Viewing should begin with the lower-power 'objective' lens. Focussing should firstly be by coarse then fine controls (if both fitted). In most microscopes, once the specimen is focussed under the low-power lens, no further focussing should be necessary if other lenses are rotated into position. CARE must be taken not to wind the high-power lens down onto the microscope slide (the slide and the lens may be damaged); focussing should be *upwards*.

The use of high-power lenses is more difficult, partly because movements in focussing or in the position of the slide need to be more delicate compared to those for low-power. For this reason, more marks may be awarded for successful use of high-power lenses (see Fig 2.12).

Fig 2.12
'Handling apparatus and materials': skill differentiation

Standard:	LEAG	MEG	NEA	SEG	WJEC	IGCSE	NISEC
High	Work in an organised way. Handle apparatus competently. Work safely.	Handle apparatus and materials correctly and confidently. Usually recognise and correct (without prompting) errors of assembly and safety	**Skills available:** *Skill 11* Assemble apparatus consisting of several components; attention must be paid to safety *Skill 12* Handle apparatus appropriately and with **care and precision** *Skill 13* Use of microscope *Skill 14* Preparing and staining of temporary mount.	Handle apparatus and materials with no error	Manipulate apparatus and/or materials correctly and safely without assistance	Handle apparatus and materials correctly and confidently. Usually recognise and correct (without prompting) errors of assembly and safety	With *minimal* supervision: Use appropriate sampling/collecting techniques for a wide range of organisms. Use a microscope with high power objective Use a wide range of biological tools and general scientific apparatus.
Mid	Work in a reasonably organised way. Usually successful in setting up and using the apparatus. Show awareness of safety precautions but sometimes careless	Use individual pieces of apparatus correctly, but have difficulties in organising equipment. Recognise and remedy errors and dangers when these are pointed out		Handle apparatus and materials with one or two minor errors	Manipulate apparatus and/or materials correctly and safely with some assistance	Use individual pieces of apparatus correctly but have difficulties in organising equipment. Recognise and remedy errors and dangers when these are pointed out.	With *supervision:* Use appropriate sampling/collecting techniques for a limited range of organisms. Use a microscope, with low power objective Use a wide range of biological tools and general apparatus
Low	Use simple apparatus. Need help with complex apparatus. Use safety precautions when reminded.	Handle apparatus and material given considerable assistance		Handle apparatus with one major error, one major error and one or two minor errors, or several minor errors	manipulate apparatus and/or materials correctly and safely only with considerable assistance	Handle apparatus and material given considerable assistance	With *constant supervision:* Use correct collecting techniques for a limited range of organisms. Use a microscope, with low power objective. Use a range of simple biological tools and general scientific apparatus.

3.4 How are marks awarded?

The 'handling apparatus and materials' skill may be assessed at the same time as that for 'following instructions' (Unit 2); the student should then 'be expected to carry out the investigation with such facility that a meaningful result is obtained' (MEG). For example, if microscope skills are inadequate, it will be difficult for the you to observe and (if necessary) record a 'meaningful' view of the specimen. Note, however, that this might be achieved with the teacher's assistance, so that you are not penalised in subsequent assessments in the same exercise, e.g. in microscope drawing.

3.5 Examples of practical work involving this skill

Most of the examples which are suggested for the 'following instructions' skill (Units 2, 3) are also suitable for 'handling apparatus and materials'. Those examples and the additional examples given below have all been suggested by Examining Groups but are by no means a complete list of possibilities:

- prepare and stain a temporary mount for microscopic examination, for example of cheek cells*, onion epidermis, (see Ch. 4, Unit 3), algae cells. *(Note: the use of cheek cells has been restricted by some LEAs, due to concern about the spread of the HIV ('AIDS') virus).
- Assemble a simple counterpoise balance, using pieces of apparatus provided, to be used to show water loss in leaves.
- Dissect out a 2 cm length of xylem from a celery petiole (previously immersed in eosin stain) using scalpel, scissors, forceps and pins.
- Following a set of instructions, investigate the structure of a heart, eye, kidney, fish gill.
- Maintain a water bath (beaker of water) over a bunsen burner at 50°C for two minutes, within 2°C above and below the set temperature.
- Set up and conduct an experiment using a 'smoking machine'.
- Make temporary mounts of rhubarb epidermis, to study the effects of different concentrations of solutions on the cell contents.

UNIT 4 OBSERVING AND MEASURING

Observation and measurement are the means by which information is obtained during a practical exercise. Even a simple exercise or experiment can provide much information. This information is only useful if it is *recorded* (and later organised and interpreted) in an appropriate way. There are two important aspects of this process:

- **Relevance:** only essential information should be recorded. Students are therefore expected to develop a capacity for *discrimination* – choosing which information to include, and which to ignore.
- **Accuracy:** observations and measurements should be as accurate as possible, within reasonable limits. This is partly determined by the techniques and apparatus being used, and also by the aims of the practical exercise.

Examples of how the sub-skills of relevance and accuracy can be used in observing and measuring are given below. Observation and measurement are *active processes;* this is also true of 'following instructions' (Unit 2) and 'handling apparatus and materials' (Unit 3). The outcome of all these practical skills may subsequently be a written account, consisting of recording and interpreting information (Units 5 and 6). Observation and measurement effectively link the 'active' and written aspects of practical work. This can therefore be an important skill area. Here observation and measurement are dealt with separately, though many practical exercises involve both.

4.1 Observation

There are several techniques involving observation in Biology. Some Examining Groups require that different forms of observation are assessed separately (Fig 2.13). Common types of observation include *description of structures, description of changes, comparison of similarities and differences* and *matching specimens*. Any of these may be used individually or in combination, according to the nature of the task. Evidence of successful observation is some form of description, which can be diagrammatic or written. In written descriptions, pupils are not normally penalised if they are unable to use correct terminology. However, the language should be clear and precise.

Fig 2.13
'Observing and measuring': skill summary

LEAG	MEG	NEA	SEG	WJEC	IGCSE	NISEC
Direct observations and comparisons made on a range of biological material and specimens. Recognise features of biological relevance.	Use standard laboratory apparatus to observe biological processes, activities and structures and to measure accurately	(see Fig. 2.25)	Make and convey accurate observations	Observe changes in an experiment. Observe structural differences or similarities. Use a key (to make accurate (and complete) identification).	Use standard laboratory apparatus, to observe biological processes, activities and structures and to measure accurately	Measure on linear (and other) scales. Estimate (accurately), select (and suggest) appropriate equipment for measuring. Observe changes in colour, form, level. Observe similarities or differences in illustrations, photographs, specimens.

DESCRIPTION OF STRUCTURES

You may be asked to examine an unfamiliar specimen and to describe distinct or indistinct features. The 'specimen' may be a living or preserved organism, a photograph or a diagram.

Unfamiliar specimens will share certain features with specimens which you have observed on other occasions; this skill is therefore a kind of comparison (see below). For example, an unfamiliar flower will have many similar features (and some different) to those of a flower examined on a previous occasion. It is, however, important to observe features that are actually present, rather than what might be expected.

Features could include shape, colour, texture and size. Structures which have a clear biological function are particularly suitable for description. Additional credit is given by some Examining Groups for descriptions of fine detail; this may require the use of a hand lens or microscope.

DESCRIPTION OF CHANGES

Many experiments in Biology involve visible changes during the experiment, for instance in the colour or clarity of solutions in food tests, in respiration (Ch. 4, Unit 13), or in some other activity. Students who are *colour blind* may have difficulty in making observations involving colour changes. Allowances are not made for such students *during the actual assessment;* instead, the school will inform the Examining Group who may then adjust the students' marks.

Observations may need to be *continuous* throughout the experiment (e.g. counting oxygen bubbles in photosynthesis, Ch. 4, Unit 16), or simply taken before and after the experiment (e.g. in a behaviour experiment).

Although observations should only be based on what is clearly seen, it helps to be generally aware of what changes might be *expected* during an experiment. This is particularly useful when changes are not immediately obvious (e.g. during growth experiments) or when the change takes place quite quickly (e.g. pulse rate before and after exercise). When changes are rapid, it is often a good idea to make brief notes during the experiment, perhaps in a table prepared beforehand. Another idea, if time and materials allow, is to repeat the experiment.

COMPARISON OF SIMILARITIES AND DIFFERENCES

> **You may be asked to compare specimens with which you are not familiar!**

Comparisons in practical Biology are often based on external features, usually quite clearly visible. Both similarities and differences in these features often reflect corresponding similarities and differences in development, function, habitat and other *characteristics* associated with the organism. If the aim of a practical exercise is 'to compare', then similarities *and* differences should be found, otherwise maximum marks may not be given. Even organisms having a very similar classification (see Ch. 4, Unit 5) will probably have obviously different features, due to variation (see Ch. 4, Unit 12). On the other hand, organisms which seem quite unrelated may share several common features.

The features which are chosen for comparison should be clearly evident, even though such features may only be fully revealed using a hand lens or microscope. It is possible that a particular feature might be due to damage or poor preparation. For example, flowers may have petals missing, preserved insects may have legs or mouthparts missing. If this problem is suspected, more than one specimen of the same type should be examined. In any case, an awareness of any unusual features is an important part of the observation process.

Comparisons should involve features which are likely to be *biologically significant;* that is, features which have some recognizable role in the organism's functioning. This will obviously require some theoretical understanding, too. For example, in a comparison of the microscopic

CHAPTER 2 OBSERVING AND MEASURING

cross-section of a shoot and root, the relative distribution of vascular tissue (xylem and phloem) and the presence or absence of root hairs are significant, but differences in colour of the sections may have resulted from the use of different stains.

It is often easier to present comparisons in a *table* (see Unit 5), with descriptions for each specimen next to a corresponding description for the same feature. This method allows a fairly systematic comparison, so that descriptions for one specimen are 'balanced' by references to another. Comparisons can also be made by separate written accounts for each specimen, or by annotated diagrams (i.e. diagrams with fully descriptive labels). Whichever method is used, it is important to emphasise equivalent structures for each specimen and to make it clear whether differences or similarities are being described.

MATCHING DIFFERENT SPECIMENS

Matching is a method by which specimens can be *identified*. Unfamiliar specimens are matched with given descriptions (e.g. in an identification key see pp 64–65) or with known specimens (e.g. photographs or labelled specimens). This is effectively a form of comparison, mainly of *similarities*.

USING IDENTIFICATION KEYS

Identification keys provide a series of (usually) paired questions or descriptions, often based on visible structures. (Note that these are selected for convenience; they may not be 'biologically significant'). These paired features progressively eliminate various possible identities until the identity of the specimen is confirmed (see Ch. 4, Unit 6). The skill of using keys has to be acquired with practice. For coursework assessment in this form of observation, students should be familiar with the use of simple prepared keys. Better still, some experience in the construction of an identification key would be useful. The use of keys may also be expected in the written examination.

MATCHING WITH TYPE SPECIMENS

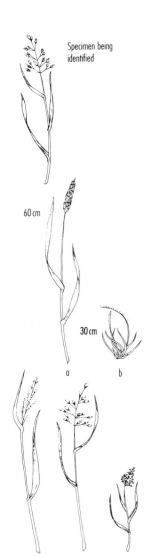

Fig 2.14
Matching with 'type' specimens

Unknown specimens can be identified by comparisons with known, or 'type', specimens. All the specimens are likely to be quite closely related. For this reason, careful examination of each will be necessary. Several (perhaps five) alternative known specimens may be given (Fig 2.14). Students may be unfamiliar with 'known' and unknown specimens, but this need not be a problem, since the object is simply to match visible similarities. Note that this is the method commonly used in natural history field guides; a plant or insect can be named by matching it against photographs or diagrams of fully described specimens.

4.2 Measurement

All Examining Groups expect candidates to be able to perform a range of measurements. Measurement is a *quantitative* process, whilst observation is qualitative. Some Examining groups also ask candidates to *estimate*; this skill is, in a sense, intermediate between those of measurement and observation.

Graduated scales are often used in measurement (e.g. in thermometers – see Fig 2.16). If the level being measured does not line up exactly with the graduation markings, it is necessary to estimate the value being measured. In effect, this means 'imagining' a more detailed scale between the actual markings (Fig 2.15). There are occasions when a scale has to be read the 'wrong way round' (e.g. in an inverted syringe); this is shown in Figure 2.15. Note in the figure that the measurement is from the base of the 'meniscus' (see below).

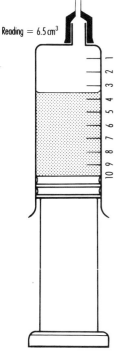

Fig 2.15
Measuring and estimating using a graduated scale (syringe)

Measurement is a means of determining quantities of temperature, time, mass, length, area and volume in biological situations. This provides information which can be used to evaluate changes during experiments, or to make comparisons between two systems. Skill in measurement is often simply a skill in using various instruments. The accuracy of measurements is partly determined by these instruments, but more importantly by the skill of the student making the measurement. Note that *units should always be given when measurements are stated*. Students often lose marks unnecessarily when they fail to do this! The units most commonly used in practical work are given below. Additional units for each type of practical work may be used in the written papers (see individual syllabuses for a complete list of units). The required degree of accuracy for each type of measurement is specified by the Examining Groups and included here.

TEMPERATURE (units: °C)

The teacher may provide a water bath of fixed temperature and students asked to make their own measurements, normally using a graduated −10°C to 110°C thermometer. The accuracy needed for maximum marks varies between Examining Groups; it may be as narrow as 0.5°C (LEAG, MEG, IGCSE), or 1°C (WJEC, NISEC). To be awarded higher marks, NEA candidates are expected to be able to use, in addition, thermometers with multiple or fractional scales, e.g. 0°C to 360°C and clinical thermometers, respectively. The graduations ('intervals') on such thermometers do not represent single units; i.e. they represent multiples or fractions of 1°C (Fig 2.16).

Measurement of temperature is particularly important in enzyme experiments (Ch. 4, Unit 2); this is so that the temperature is known and also so that fluctuations in temperature can be controlled. Temperature is also important in experiments to study cooling effects (Fig 2.17).

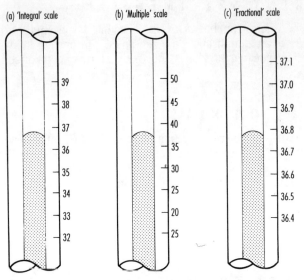

Fig 2.16
Comparison of integral, multiple and fractional graduated thermometers all reading the same temperature

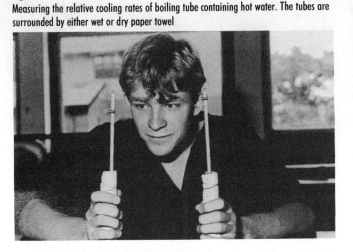

Fig 2.17
Measuring the relative cooling rates of boiling tube containing hot water. The tubes are surrounded by either wet or dry paper towel

TIME (units: s/h)

Students are asked to measure *time* to within 1 second (s) accuracy. Timing is usually undertaken with a stopclock or stopwatch. Some biological 'events' can be quite brief, so accuracy is often determined by the speed at which the stopwatch or clock is turned on or off! Precision can therefore be improved by familiarity with the controls, and also by anticipating the beginning and end of the event being timed. This is made easier when students are working in small groups, perhaps with one making observations or recording information whilst another operates the timing device.

In human physiology experiments, the human subject should preferably not be involved in timing because this can affect the result. Examples of such experiments include the measurement of pulse rate (Ch. 4, Unit 14) and reaction time.

> **If you are working in a group, it is a good idea to swap tasks during the exercise so that each student has an opportunity to develop different skills**

MASS (units: g/kg)

Mass is commonly determined by the use of various kinds of balance. Many balances give a direct digital reading to the nearest 0.1 g; this is within the accuracy limit of 0.5 g (or even 1 g) specified by Examining Groups. However, all balances need to be operated properly. For instance, the balance should read zero before weighing begins; this can be done (e.g. by the teacher) by the use of a 'tare' control. Also, students need to take into account the mass of the container holding the substance being weighed. The mass of the 'container' (which might simply be a piece of paper) should be determined *before* the object itself is weighed. The first mass (container) can then be subtracted from the second (container + object) to obtain the true mass of the object. Some balances can be zeroed so that a direct reading of the object's mass is given, without the need for any calculation.

As with all measuring instruments, an important part of the skill is familiarity with their operation. Balances are among the most expensive types of apparatus in the Biology laboratory; they are sensitive and therefore easily damaged by incorrect use. Objects (especially if they are wet) should not be placed directly onto the balance 'pan'. In particular, no attempt should be made to weigh hot objects.

LENGTH (units: mm/cm/m)

The unit used for measurement of *length* will be determined by the scale of the object being measured. For instance, distances along a transect in ecology will probably be in metres (m), the heights of seedlings (Ch. 4, Unit 9) are best expressed in cm (or perhaps mm) and the length of a leaf in mm (Fig 2.18). So, the choice of suitable units is one aspect of this skill. Accuracy becomes more important in measuring small distances, since even small errors can make a dramatic difference. For higher marks, Examining Groups accept length measurements within 1 mm (0.1 cm).

Fig 2.18
Measuring the length of a leaf

Fig 2.19
Examining leaf area

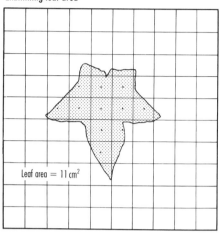

A R E A   (units: $mm^2/cm^2/m^2$)

Area is a 2-dimensional measurement and can be determined directly in regular-shaped objects by multiplying length and width. However, in Biology, shapes are often irregular and area is *estimated*, often using some form of regular grid for reference. For instance, the area of leaves can be found by tracing the outline onto graph paper. The total number of complete squares within the outline is then determined by counting and estimating (Fig 2.19).

A similar method is used for finding the area occupied by different types of vegetation within a quadrat (Fig 2.20;). The areas of 'percentage cover' are determined by estimation (Fig 2.21), though some measurement may be used. Quadrats may be subdivided into grids for more accurate estimates (see Ch. 4, Unit 18 for an example of this).

Fig 2.20
Using a quadrat to estimate percentage cover

Fig 2.21
Diagram showing percentage cover

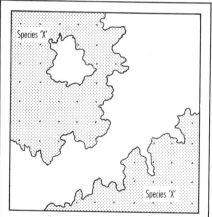

V O L U M E   (units: cm^3/m^3)

In Biology, *volume* is most commonly used in measuring liquids. There are several simple pieces of apparatus such as the measuring cylinder, syringe (see Fig 2.15) and (less commonly at GCSE) graduated or bulb pipettes or gas burettes. The volume markings on beakers are generally unsuitable for accurate volume measurements.

Examining Groups ask for accuracy up to 0.5 cm^3. For NEA, the accuracy limit is 1 cm^3, e.g. using a 250 cm^3 measuring cylinder. This skill is clearly more demanding when the graduations of the measuring device do not represent whole units; for instance in 0 to 10 cm^3 and 0 to 500 cm^3 measuring cylinders (this is similar to the situation in thermometers – see Fig 2.16).

The correct size of measuring cylinder for a particular experiment may be provided, or specified in the instructions. If this is not the case, it is important to use an appropriate size. For instance, 15 cm³ of glucose solution should be measured in a 25 cm³ measuring cylinder (if available); alternatively, a 10 cm³ cylinder could be used to measure twice, e.g. 10 + 5 = 15 cm³. A 500 cm³ or 1 000 cm³ (1 dm³ or litre) cylinder would be too big for this measurement (and the scale may not begin at 0!).

Students may need to measure a very precise amount of liquid into a measuring cylinder; this can be easier if a pasteur pipette is used (especially when the required amount is almost reached). Note that all liquids should be measured at the base of the meniscus, not from the part of the liquid at the edge which 'climbs' slightly up the container.

The volume of solids may need to be measured in Biology, too. Volume can be calculated for regular-shaped objects by the product of length × width × height or, more commonly, length × cross-sectional area. This can be used to measure cylinders (e.g. see Ch. 4, Unit 4); the volume of a cylinder is πr^2 × length.

However, many objects in Biology are irregular in shape; the volume of such objects, if submersible, can be found by the displacement of water. Such objects (e.g. a batch of seeds), when placed in a measuring cylinder containing water, will cause the water level to rise. The amount by which the level rises is equal to the volume of the object being immersed.

The different types of measurements described above will be used in various ways throughout the Biology course. The teacher may organise a coursework assessment specifically in measurement skills. These may be assessed separately, or in combination. For example, heating a known amount of water for a set length of time and measuring temperature before and after heating will required three types of measurement.

> **Measuring skills can be practised at regular intervals during the Biology course**

There are two further measurement skills which are often required in biology:

SCALING

Diagrams or photographs of specimens are produced to a convenient size. For example, Examining Groups may ask that students draw diagrams to occupy a space at least half A4 (i.e. about 300 cm²). Some specimens can be shown as their actual size, i.e. *life size*. However, some specimens will be either too small or too large to be shown conveniently in this way. Clearly, this will involve enlarging some objects and reducing others. It is therefore useful to be able to express the size of a diagram or photograph (the *representation*) in terms of its *actual* size, i.e. to use a *scale*.

The relative size of the representation and the actual specimen is the *scale factor*, often shown by an '×' symbol. Scales are usually based on convenient increases (or decreases), e.g. '× 10', '× 2' (increases), '× 0.1' (decrease). Scale is often based on length, and may also be expressed as a ratio, e.g. 10:1 magnification; the scale factor is × 10. You may also be asked to interpret a given scale.

There are two important steps in constructing a scale (see below) or interpreting a scale:

▶ Check that the *same units* are being used for the representation and the specimen *before* any calculation is done. (More convenient units can be used once a value has been obtained).
▶ Decide whether the calculation is a *multiplication* or a *division* by the scale factor. This depends on the relative sizes of the specimen and its representation. It also depends on the way in which the scale factor is expressed. For example, multiplication by '0.1' will have the same effect as division by '10'! All this can be confusing, but with practice the figures can be handled more confidently!

The use of scales can be made clearer with some examples:

Example 1: Scaling down

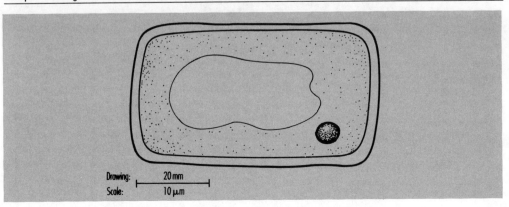

Fig 2.22
Drawing of cell, with scale

In the drawing of a cell (Fig 2.22) the scale is shown. Each 20 mm on the drawing represent 10 μm on the *actual* cell. The unit μm is used to measure small (i.e. microscopic) lengths; 1 μm = 0.001 mm (1 μm is one-thousandth of a mm). Using this information, the *scale factor* of the drawing compared with the actual cell can be calculated:

▸ *Convert the scale to the same units:*
  10 μm = 0.01 mm   [i.e. 10 × 0.001]
▸ *Express the size of the drawing in terms of the actual size:*

$$\frac{\text{drawing}}{\text{actual}} = \frac{20}{0.01} = 2000 \quad \bigg| \quad \text{The scale factor is therefore} \times 2000, \text{ or } 1:2000$$

> You may find this difficult at first. If necessary, ask the teacher to go through these ideas with you

Using this scale factor, the actual length of the cell can be calculated, since any dimension on the drawing is × 2000 its actual size in the cell:

▸ *Length of the cell in the drawing:*
  The length of the cell in the diagram is (approximately) *68 mm*.
▸ *Length of the actual cell:*

$$\frac{\text{measured length}}{\text{scale factor}} = \frac{68 \text{ mm}}{2000} = 0.034 \text{ mm} = 34 \text{ μm}$$

In GCSE Biology you may be required to perform this sort of calculation in the written examination, depending on the Examining Group, or the Paper being taken. In *coursework*, specimens under the microscope obviously cannot be measured directly with a ruler. However, students *are* expected to give the scale of their microscope drawings. There are two main ways of doing this:

> Examiners will often give extra marks for correct use of scales

a) **Magnification.** The size of the specimen depends on *combined magnification of the lenses being used*. This can be calculated from the eye piece lens and the objective lens being used (see Fig 2.8). If a specimen was being examined using a × *10 eyepiece* and a × *40 objective* lens, the magnification of any specimen will be 10 × 40 = 400 times its actual size.

However, the *drawing* of the specimen may be bigger or smaller than this! For instance, the cell shown in Figure 2.22 is drawn about 5 times larger than it would actually appear down the microscope using the above lenses. So, its magnification is actually 400 × 5 = 2000 times. Students may find this idea difficult at first, but they will gain additional marks by giving the correct magnification.

b) **Measurements using calibrated scales.** It is possible to give the scale of a microscopic specimen by using a small scale which fits into the eyepiece. This piece of apparatus is not used by many schools at GCSE level, and is a skill expected only of more able NEA candidates.

Example 2: Scaling up

Fig 2.23
Photograph of a skull

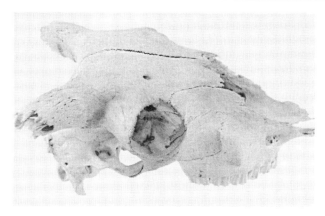

In this photograph of a skull (Fig 2.23), the *scale factor* is 1:4; the skull is '× 4' larger than its photograph. Supposing the length of the actual skull 'A' is to be found:
Length of skull in photograph = 50 mm.
Therefore length of actual skull = 50 × 4 = **200 mm** (= 20 cm).

SYSTEMATIC COUNTING

There are occasions in practical Biology when it is necessary to count relatively large numbers (e.g. more than 100) of similar structures or similar events. There is a possibility of omissions or duplicated counts if care is not taken. This can be avoided if the counting is organised. This often means breaking the task down into smaller units, and/or sharing the task within a group of students. There are two aspects to systematic counting:

Observations/measurements

The objects being counted can be divided into sub-groups, for example, using:
- *grids* (real or 'imagined') e.g. quadrats with internal divisions (Fig 2.24)
- *arrays* e.g. lines of privet leaves (Ch. 4, Unit 12).

Fig 2.24
Sub-dividing a quadrat using a metre rule

Recording

A commonly-used method is to make marks on paper which can be added up later. For instance, dots on paper to record bubbles of gas given off by an aquatic plant (Ch. 4, Unit 16), since the events can be very rapid.

The *tally* is an effective technique for recording a large running total. Tally counting involves counting in groups of *five*, which can then be easily added up. A completed group is shown by a diagonal line (the fifth line). For example, a count of 23 petals could be recorded like this:

⌿⌿⌿⌿ ⌿⌿⌿⌿ ⌿⌿⌿⌿ ⌿⌿⌿⌿ III

Hand-held mechanical *tally counters* may be available to make things easier. Each depression of a button at the top causes the total to increase by one. One example of the use of this might be to count the colonies of bacteria on an agar plate.

4.3 How are marks awarded?

Fig 2.25
'Observation and measurement': skill differentiation

Standard:	LEAG	MEG	NEA	SEG	WJEC	IGCSE	NISEC
High	*Comparisons:* Suitable features selected, relevant points clearly identified. Chosen features accurately described in a suitable form. *Measurements:* All measurements accurate	Handle quantitative work with confidence. Use measuring apparatus correctly and relate readings to quantities. Make detailed and accurate observations.	**Skills available:** *measurement:* Skill 1 Reading scales (Whole number scales) Skill 2 Reading scales (multiple/fractional scales) Skill 3 Measuring out quantities Skill 4 Scaling Skill 5 Systematic counting *observation:* Skill 6 Matching Skill 7 Describing (gross details) Skill 8 Describing (fine details) Skill 9 Finding differences (gross features) Skill 10 Finding differences (fine details)	Convey observations with complete accuracy	*Observation:* Observe very accurately and give complete sequence of changes. Observe a minimum of three structural differences/similarities. Use a key to complete an accurate identification. *Measurement:* Accurate within ± 1.0 mm, 1.0 cm³, 1 sec, 1°C	Handle quantitative work with confidence. Use measuring apparatus correctly and relate readings to quantities. Make detailed and accurate observations	Decide independently which observations and measurements are most appropriate. Measure from a wide range of scales. Estimate accurately. Observe and describe many changes, and similarities and differences without guidance.
Mid	*Comparisons:* Suitable features usually selected. Chosen features described reasonably well *Measurements:* About half of the measurements accurate	Qualitative observations usually accurate in main features, but quantitative observations lack detail and discrimination		Convey observations with one or two minor inaccuracies	*Observation:* Observe more than one, but not complete sequence of changes. Observe two structures similarities or differences. Use keys but unable to complete accurate identification *Measurement:* At least *two* measurements accurate within above limits	Qualitative observations usually accurate in main features, but quantitative observations lack detail and discrimination	Provided with details of observations and measurements to be made. Measure from a range of scales. Estimate accurately. Observe and describe several changes, and many similarities or differences.
Low	*Comparisons:* Suitable features chosen with difficulty. Chosen features poorly described. *Measurements:* (see Fig. 2.40)	Make detailed and relevant observations and measurements given considerable assistance		Convey observations with one major inaccuracy, one major inaccuracy and one or two minor inaccuracies or several minor inaccuracies	*Observations* Poor observation. Observe only one structural similarity or difference. Follow only the initial stages of a simple key. *Measurement:* one of the measurements accurate within above limits	Make detailed and relevant observations and measurements given considerable assistance.	Given observations or measurements which must be made. (Measure a limited range of scales) Observe and describe *one* change, and two similarities or differences.

CHAPTER 2 **RECORDING AND COMMUNICATION** 23

4.4 Examples of practical work involving this skill

There are many examples suggested by Examining Groups of practical exercises involving observations and measurement. For convenience, practical work involving mostly observation are listed separately from work mainly based on measurement:

Observation

- Identification of similarities/differences between an insect and a millipede.
- Comparison of an earthworm and an arthropod.
- Find differences between two species of woodlouse.
- Identify cell types in a microscopic section.
- Construct and use a key.
- Identify a specimen using a simple key.
- Observe in a wallflower the number, colour and relative position of sepals, petals, stamens and carpel.
- Find the differences between a prepared stem section and root section.
- Find differences between two photomicrographs of plant or animal cells.
- Describe, in a small freshwater fish, eyes, mouth, operculum, number and position of gills and overall shape.
- Compare inhaled/exhaled air.
- Make accurate observations of flowers of different species, using a hand lens.
- Make a detailed observation of the range of results produced by a series of food tests.
- Compare an artery and vein, in transverse section, using a microscope.
- Identify a specimen, using a simple pictorial key.
- Determine the pH of a range of different soils.

Measurement

- Systematic count of coloured maize grains showing inheritance.
- Determine the nematode population in an earthworm.
- Systematic count of prickles on leaves.
- Measure the expansion of yeast dough.
- Investigate changes of mass of potato chips in a series of sugar solutions.
- Measure rates of water uptake by a plant, using a potometer with graduated scale.
- Determine the number of stomata on a given area of leaf.
- Determine genetic ratios of grains on a corn cob.

UNIT 5 RECORDING AND COMMUNICATION

Recording and communication is an important link between observation and measuring (Unit 4) and interpretation (Unit 6). Examining Groups expect you to record and communicate information with *accuracy* and *clarity*. This is important to the student who conducted the practical work, because it is a means of *organising* the information so that it is meaningful. Accuracy and clarity is also important to teachers and examiners, who were not directly involved with the work.

Fig 2.26
Recording and communication: skill summary

LEAG	MEG	NEA	SEG	WJEC	IGCSE	NISEC
Drawing: Recognise features of biological relevance and record and communicate the findings in an appropriate manner. Record results from conducting experiments.	Record relevant data accurately and clearly. Data may take the form of tables, graphs, diagrams, drawings and written accounts.	(see Fig 2.40)	Record results in an orderly manner	Record changes during an experiment. Record similarities and differences between two specimens. Record correct proportions and size.	Record relevant data accurately and clearly. Data may take the form of tables, graphs, diagrams, drawings and written accounts.	Describe or report a biological procedure or experiment, verbally or in writing. Record information in an appropriate form.

5.1 Accuracy

Information collected by observation and measurement should be *precise* and *relevant*, within reasonable limits.

ACCURACY LEVEL

Examining Groups specify the minimum acceptable accuracy (for maximum marks – see Unit 4.2). However, accuracy level is also determined by the purpose of the experiment and by the limitations of the apparatus used. For example, if differences in area between different nettle leaves is as much as 20 cm^2, then measurements to the nearest 1 mm^2 will be unnecessary and also time-consuming. In any case, the method used (see Fig 2.19) is partly based on estimation, and would not be expected to give such precise values.

Some apparatus can give measurements which are *too* accurate! For instance, a top-loading balance might give a reading to the nearest 0.001 g (i.e. 1 mg). This is clearly too precise for an experiment designed to show progressive loss in mass; the precise mass at any given time is not particularly important. Examining Groups specify an accuracy limit of 0.5 g for mass. So, a measured value of 20.383 g can be recorded as 20.5 g.

EXPRESSION OF ACCURACY

Decimal places

Accuracy can be expressed to a given number of decimal places. (Another method, 'significant figures', is less often used in Biology and is not covered here.) Accuracy is conveniently expressed in terms of the decimal point, especially if the value involves a 'fractional' part (i.e. less than one). For example, 40.2°C, 12.5 s, 141.6 cm^2, 0.7 g are all expressed *to one decimal point* (sometimes written as 'to 1 d.p.'). This is the maximum level of accuracy (for the smallest units) in GCSE Biology. If a *more* accurate measurement is made, this can be expressed accordingly by 'rounding up or down' additional figures. The accepted rule is to *round up* figures 5 and over (i.e. 5, 6, 7, 8, 9) and to *round down* figures below 5 (i.e. 4, 3, 2, 1, 0).

In the following three examples, each of the measured values (given to two or three decimal places) is expressed to one decimal place:

40.86 g → 40.9 g; 38.94°C → 38.9°C; 9.245 cm^2 → 9.3 cm^2.

Although measurements may be made at an appropriate level of accuracy, *calculations* can result in a final value which seems to be *more* accurate! For example, the lengths of six privet leaves (see Ch. 4, Unit 12) were:

2.4, 3.1, 3.7, 2.9, 2.2, 3.6 (all values are in cm.)

The mean ('average') of this data is 2.9833333333333 (etc) cm. This implies that measurements were made to a very high level of accuracy, which was not true! So the mean value can more conveniently be expressed to one decimal place: 3.0 cm. In general, the results of any calculation should not be expressed at an accuracy level above that of the least accurate measurement.

Experimental error

Rather surprisingly perhaps, experimental error is fully acceptable to both teachers and examiners! This 'error' is not necessarily caused by poor experimental method; it often results from *known* and *expected limitations* of the apparatus or technique being used.

Students may simply be required to have an awareness of the variability and limitations in the experimental procedure. Experimental error within an experiment is made apparent by the use of *repeated measurements* in the *same situation*. These should, in theory, be the same. Any differences which occur are evidence of experimental error. Measurements of the same situation can be repeated by a single student or group of students, or by combining class results. It is, however, important that the same (or very similar) conditions are used.

For example, consider an experiment in which cylinders of potato were immersed in 5 per cent sucrose for 15 min (see Ch. 4, Unit 4). The decrease in length of potato cylinders from eight different containers was as follows:

0.3, 0.5, 0.2, 0.3, 0.4, 0.4, 0.3, 0.4 (all values are cm).

The mean value = 0.4 cm (to 1 decimal place)

As a value, experimental error can be expressed in terms of maximum difference (plus or minus) from the mean; in this case the experimental error is ± 0.2 cm, i.e. the difference between 0.2 (the furthest reading from the mean) and 0.4 (the mean).

Revealing experimental error in this way can be useful as students may be expected to identify possible *sources of error* as part of interpretation (Unit 6). In this example, the most likely sources of error are:

- 5 per cent solutions not all made to same concentration
- different amounts of 5 per cent solution used
- potato cylinders not all fully immersed
- variations in potato tissue (e.g. from different potatoes)
- incorrect measurements (e.g. volume, time, length)
- incorrect calculations.

Some of these errors are more difficult to avoid than others. Even with a more careful approach, repeated measurements may still show differences. However, the experiment may remain valid because differences *between* treatments are greater than differences *within* treatments.

Relevance

An important 'hidden' skill in recording and communicating is knowing how much of the available information to include, and how much to omit. Experiments often generate more information than can be used. This means that some *selection* of information will be necessary. This process will also occur at the observation and measurement stage (Unit 4) and during interpretation (Unit 6). Selection of information requires practice and confidence. Too little information can result in an incomplete or misleading interpretation. Too much information can be very time-consuming and may obscure the more important aspects of the investigations. Knowing what information is relevant comes from *understanding the purpose* of the experiment or exercise.

Many investigations in Biology involve some form of comparison, for example between two specimens (e.g. by drawing or listing comparable features). Comparisons also occur between different situations (e.g. the relative activity of a tissue or organism in changing conditions). The information which is used should be relevant to the comparison. For instance, in a drawing showing two different insects (see Ch. 4, Unit 5), sufficient information should be included to emphasise the main differences and similarities.

In an experiment to show the sensitivity of maggots to light and dark the relative numbers of maggots in light and dark regions of the tube are counted after a given time. The rate of their movement *may* be relevant, but this is dependent on other factors, such as temperature.

In Biology experiments where several comparisons or measurements are involved, the most useful information will probably indicate a *trend*. A trend is an overall pattern of results. For instance, repeated measurements of leaf length in privet (Ch. 4, Unit 12) reveals variation, with most leaves having an intermediate length. Recording differences in shape or colouration, or even width, thickness and area, would not necessarily provide further evidence for this trend, though each of these might reveal very useful information about other trends! Each of these trends may or may not correspond to the trend for length, so a decision needs to be made about the time available for measurements.

5.2 Presentation of results

One way to limit the amount of information included is to remember that it must all be presented in a suitable way and then interpreted (Unit 6). The remainder of this Unit describes some of the techniques needed for the appropriate presentation of results. The most common methods are *diagrams, drawings, graphs, tables,* and *written accounts*.

DIAGRAMS

> **Practise drawings and diagrams in rough first if you do not feel confident of your ability**

It is perhaps necessary to make a distinction between diagrams and drawings in Biology. *Diagrams* are used to illustrate the method and results of an experiment or exercise. Their main purpose is often to show the relative positions of different pieces of apparatus or materials. For this reason, diagrams should be highly simplified and need not conform to a scale. *Drawings* (see below), on the other hand, often show a considerable amount of detail, with all parts in the correct proportion to each other. In practical Biology, diagrams tend to be of a 'situation', whilst drawings are usually of specimens. For further explanation, refer to the different examples of diagrams and drawings at the end of this Unit.

The main skill in constructing diagrams is a *recording* skill. Diagrams should be used as a means of *explaining* something, and can save a lot of writing. (This applies in the written examination, too!) Learning how to construct diagrams rapidly is a very useful skill.

Diagrams will, in fact, be needed in most practical write-ups in Biology, so you will be given many opportunities to practise. In any case, some pieces of apparatus are fairly standard in many experiments. For example, test tubes, boiling tubes, 250 cm^3 beakers and thermometers occur in many experiments. Bunsen burners and tripods are common too although they do not necessarily need to be included (see Figs 2.27 and 2.28 below).

Diagrams should be *two-dimensional* (i.e. 'flat'), showing only the outline and any important contents. You should resist the temptation to draw in three-dimensional diagrams; they are difficult to draw (a tripod is *very* difficult!), time-consuming and may in any case be confusing. You should, if possible, also avoid using stencils for scientific diagrams. The problem with stencils is that they limit the opportunity to learn the skill of 'free-hand' drawing. Also, the relative sizes of different types of apparatus may be unsuitable for a given diagram. A further problem, frequently encountered by students, is getting glass tubes to connect up pieces of apparatus successfully!

Figure 2.27 shows a photograph of a fairly typical set of apparatus in Biology. Compare this with the diagram of the same apparatus (Fig 2.28). Notice that the diagram omits much of the detail in the photograph and emphasises important features. Labels should be brief, clear and should not obstruct the diagram. You can avoid writing over label lines by putting each label close to the structure that it refers to.

Fig 2.27
Photograph of apparatus used in an experiment

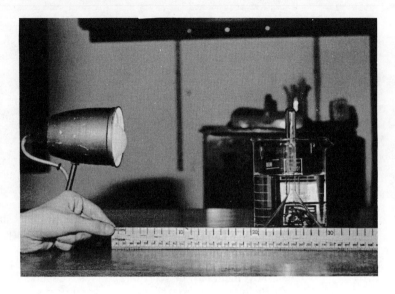

Fig 2.28
Diagram of the apparatus shown in Fig 2.27

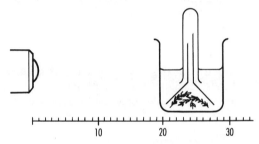

Figure 2.29 again shows the apparatus pictured in Figure 2.27. However, Fig 2.29 includes many of the mistakes which are often made with this sort of diagram.

Fig 2.29
Diagram of the apparatus shown in Fig 2.27 – common mistakes are shown

❝ Note that whilst this diagram 'looks' impressive, it was time consuming to draw, is too detailed and contains inaccuracies ❞

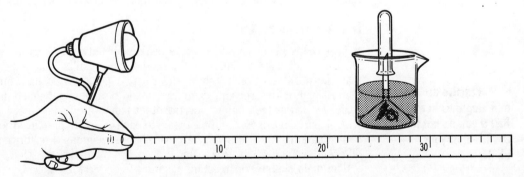

DRAWINGS

The main skill required for biological drawing is *observation* (see Unit 4.1) of a specimen. In a sense, the drawing is evidence that the student has made a series of relevant observations. In particular, drawings can show that the student has an awareness of *detail* and *proportion*.

Drawings should ideally be *large* (at least half an A4 page) and drawn on blank (not lined) paper. Suitable pencils are 2H or HB; and outline of the specimen can be very lightly drawn whilst the proportions and detail are being established. In the finished drawing, lines should be clear and distinct, not 'sketched'. The three drawings (Fig 2.30 a–c) are of the same specimen, but are drawn to different standards. Compare the three drawings for size, detail, proportion and the quality of the lines.

Fig 2.30
Three different standards of drawing of the same specimen: (a) high standard (b) moderate standard (c) low standard

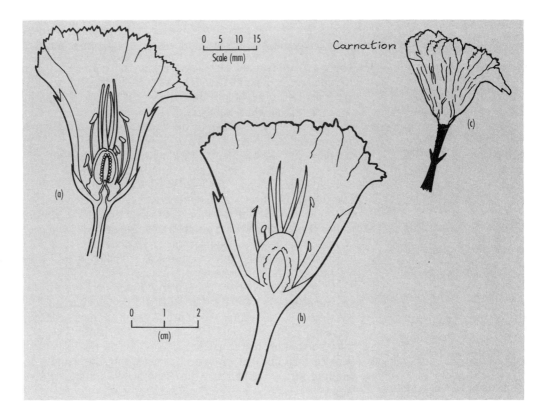

The drawing should emphasise 'biologically-important aspects' (LEAG) of the specimen. However, only aspects which are actually *visible* on the specimen should be included, otherwise the drawing will look like a thinly-disguised textbook diagram! Structures which are expected may not always be present; if not, they should not be drawn in.

Some Examining Groups (e.g. LEAG) do not require drawings to be labelled. Labelling is really another skill ('identification') which tends to be examined more in the written examinations. If the pupil is instructed to provide labels, these should be done neatly and should not obscure the drawing.

GRAPHS

Graphs are 'visual summaries' of data. Graphs can show trends in data which are often not easily apparent in the figures themselves. Candidates in GCSE Biology will be given many opportunities to demonstrate graph skills. In coursework, this will mainly consist of constructing graphs. In written examinations, you will be required to construct and interpret graphs. There are two types of graph commonly used in Biology; *line graphs* and *histograms/bar charts*. You may also be expected to construct (or interpret) *pie diagrams* or *kite diagrams*. The use of each of these four forms of graph is described below:

Line graphs

Fig 2.31
Example of a line graph: changing number of leaves on a seedling with time

❝ Graph skills are very important in Biology ❞

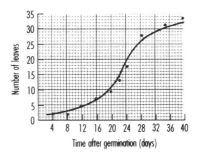

Line graphs (sometimes called 'jagged line' graphs) are used to show *continuous data*, i.e. data which changes in a progressive way. For example, the number of leaves on a seedling during a certain time interval. The measurements were as follows:

Time after germination (days):	8	12	16	20	22	24	26	28	34	40
Number of leaves present:	2	4	7	9	14	18	23	28	31	34

The data is not very informative expressed in this way; a line graph should reveal any trends. With any line graph, the following points should be remembered:

▶ The *x-axis* (horizontal scale) is used for the *independent variable* (sometimes called the 'manipulated independent'). These are the measurements which are decided by the experimenter, often before the experiment actually begins. In the example above, it is the time intervals at which numbers of leaves are counted.
▶ The *y-axis* (vertical scale) is used for the *dependent variable* (sometimes called the 'responding dependent'). These dependent measurements are the outcome of the independent measurements. In a sense, they are 'decided' by the organism within the experiment! In the example above, this is the number of leaves counted.
▶ Suitable divisions should be used in each scale so that the complete graph fills most of the available space. This will be dependent both on the type of graph paper being used and on the range of data. For instance, if the experiment above was continued for another much longer period and a similar graph drawn, the divisions would need to allow for this, on both scales. Scales need not begin from zero if all the data have values much more than this.
▶ Clear labels should be given for the scales (include units – e.g. 'days' – where appropriate) and title.
▶ Short straight lines can be used to *connect up points* (coordinates), if there is a fairly smooth change in the data. The line can be gently curved if there is a significant change in 'steepness' of the points. When points are uneven or scattered it is better to draw a line or curve of 'best fit'; this line will pass closely to as many points as possible, with roughly the same number of points on either side of the line.
▶ If more than one set of data is drawn using the same axis, each line must be clearly different (e.g. in colour or type), and should be clearly labelled (directly or by using a 'key').

> **Practise this technique, e.g. by using data from Ch. 4 of this book**

Histograms/bar charts

Histograms and bar charts are used to display *discontinuous*, grouped data. Each group of data is related to the other groups, but there is no progressive change as with data plotted in a line graph. For example, the distribution of leaf number within a plant population (Fig 2.32).

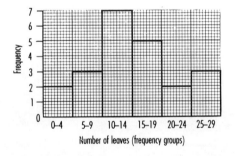

Fig 2.32
Example of a histogram: distribution of leaf number within a plant population

In *histograms*, the groups are shown as intervals along the x-axis; there is no 'overlap' between groups so any value can only belong to one group. The number of values in each group, or *frequency*, is shown on the y-axis. Such graphs are sometimes known as *frequency histograms*. Data may need to be prepared before it can be represented by a histogram. The data for Fig 2.32 are shown below:
Leaf numbers in 22 plants:
6, 15, 21, 6, 16, 18, 25, 26, 17, 12, 2, 15, 4, 14, 12, 10, 12, 20, 25, 7, 12, 11
The procedure for organising data in this form is as follows:

▶ Find the minimum and maximum values, i.e. the *range*. In this case, it is 2–26. This will determine the extent of the x-axis – in this case the scale will be about 0 to about 30.
▶ Decide on a convenient size for the *frequency groups* (x-axis), so that all data can be placed into about five or six groups. The groups should be equal in size. In this example, the range 2–26 has been broken up into six groups. Each group represents a sub-range of five values, e.g. 0–4 (i.e., 0, 1, 2, 3, 4), 5–9, 10–14 etc. Notice that any particular value can only belong to one group. This would not be so if there was 'overlap' between the groups, e.g. 0–5, 5–10, 10–15 etc. (Which groups should the values 5 and 10 be put into?) In some experiments, the data may not all consist of whole numbers; the x-axis groups would have to allow for this, e.g. 0.0–4.9, 5.0–9.9, 10.0–14.9 etc.

- Place each value from the original data into its correct group. This can be done using a *frequency table,* so-called because it allows the number of values in each group (the frequency) to be determined (Fig 2.33). Each value can be crossed out on the original data list once it has been included in the table. A tally method (see Unit 4.2, p. 22) can be used to add up data in each group.
- Plot frequency values against the y-axis for each group. Any group which is 'empty' (i.e. with no values in it) has a frequency = 0. This should be shown as a blank on the graph (see Fig 2.35)

Fig 2.33
Frequency table for the histogram shown in Fig 2.32

x-axis: leaf number groups	0–4	5–9	10–14	15–19	20–24	25–29
y-axis frequency	11 = 2	111 = 3	ⅬⅡ 11 = 7	ⅬⅡ = 5	11 = 2	111 = 3

Histograms give a fairly immediate impression of the distribution of data. However, information is 'lost' if the size of each group is too large. In the example opposite, the overall trend of leaf number would not be so obvious if the groups were much larger (Fig 2.34). On the other hand, if the groups are too small, it is difficult to observe any trend (Fig 2.35).

Fig 2.34
Example of a histogram using frequency groups which are too large

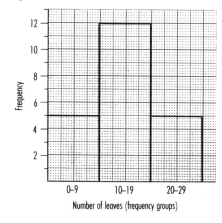

Fig 2.35
Example of a histogram using frequency groups which are too small

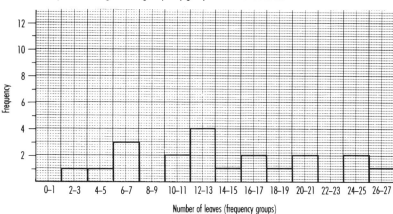

Bar charts are rather similar to histograms, but the frequency groups are non-numerical. The x-axis therefore does not represent a scale of changing values. Each group tends to be separated from the next by a gap; this shows that each 'bar' represents an isolated ('discrete') set of data. For example, the distribution of the powdery green alga *Pleurococcus* on four different surfaces of a tree trunk (see Ch. 4, Unit 18) (Fig 2.36).

Fig 2.36
Example of a bar chart: distribution of *Pleurocococcus* on different sides of a tree trunk

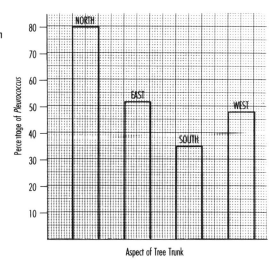

Pie diagrams

Pie diagrams are a means of representing different proportions of a given quantity as sectors ('segments') in a circle (Fig 2.37). The size of each sector is measured in degrees (°). The total number of degrees 'contained' in any circle is 360. This means that the angle of each sector (at the centre of the circle) must be calculated as a proportion of 360°. The formula for this calculation is:

$$\frac{\text{quantity} \times 360}{\text{total quantity}} = \text{angle of sector (°)}$$

It is not easy to measure fractional parts of degrees (i.e. values which are not whole numbers), so some approximation may be necessary. It is worth remembering that the pie diagram is not necessarily intended to be an accurate type of graph; its main purpose is 'visual impact'.

For example, a pie diagram could be used to display the results of our survey showing the distribution of *Pleurococcus* on a tree trunk. The data are shown in Fig 2.36a. The angle of each of the four sectors has been calculated, using the formula above:

Fig 2.36a

'Aspect' of tree trunk	% cover of *Pleurococcus*	Angle of sector (°)	Approx. angle of sector (°)
North	80	134.0	134
East	52	87.1	87
South	35	58.6	59
West	48	80.4	80
totals:	215	360.1	360

The calculated angles are initially expressed to one decimal place. These angles would be difficult to measure (and they don't add up to 360°C!) so they are approximated. Remember that the original data were, in any case, merely estimated. The total of angles of all sectors must be 360°. This is a good way of checking the calculations. It is often easier to begin with the largest angle (in this case, the angle for 'North') and working 'downwards'; the last angle does not need to be actually measured!

The resulting pie diagram for this example is shown in Figure 2.37. The sectors can be coloured or shaded to make the diagram more effective. Each sector should be clearly labelled, either directly or by using a key. The diagram should also have a title, and perhaps some further explanation.

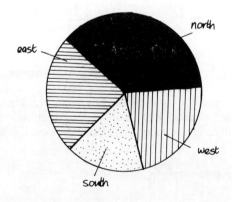

Fig 2.37
Example of a pie diagram: distribution of *Pleurococcus* on a tree trunk

Kite diagrams

Kite diagrams are commonly used to show continuously varying data from studies in ecology. For example, the relative numbers of different organisms can be determined at regular intervals along a line, or *transect*. The changing patterns in the distribution of the organisms results in a *profile* of the study area (Fig 2.38). This can include both biotic factors (organisms) and also abiotic factors (e.g. soil pH, light). Comparisons of kite diagrams drawn for each of the measured factors can reveal interactions between them.

For example, a kite diagram can show changing distribution of different plants across a footpath. The relative amounts of each plant can be estimated as a percentage cover within a quadrat. The percentage cover for two species is shown below. The quadrats were placed at regular intervals (indicated by the scale) along a line transect:

quadrat number	1	2	3	4	5	6	7	8
Species 'A'	24	35	15	8	5		8	40	20
Species 'B'	55	20	6	0	0	10	30	60

Notice that the percentages do not total 100 in any of the quadrats. This is because there are other species (also bare ground) which have not been included in the survey. There *does* seem to be a distinct pattern for each of Species A and B in this data. This can be confirmed by representing the data in a kite diagram.

Separate sets of data (in this example, data for Species A and B) are usually represented on different kite diagrams (see Fig 2.38). Each kite diagram is constructed around a horizontal base line; this represents a 'transect scale'. Values for each quadrat position are represented on *both* sides of the base line, at equal distances from it. This means that each value is plotted by two points. The total distance between the two points is determined by the vertical 'percentage scale'. This vertical scale is really two halves of a single scale. This may seem unnecessarily complicated, but it will allow a symmetrical diagram to be produced.

Each value needs to be plotted on either side of the base line, in each case as two half-values (on a 0–50 scale). For example, for Species A, quadrat 1, the value is plotted as '12' on each side of the base line. Odd numbers are more 'messy' when halved; e.g. '35' is represented by two lots of '17.5'. It is, however, sufficient to plot points approximately. Like the pie diagram (see above), the kite diagram is not intended to be a highly accurate type of graph. Also, remember that the percentage cover values are estimations, so are not likely to be very accurate themselves.

Next, for each kite diagram, all the points either above or below the base line are connected with each other by straight lines. This results in a complete outline for the kite diagram itself. The outline can then be shaded or coloured in. Figure 2.38 shows a complete and also a partially completed diagram for the above data. Notice that the two kite diagrams are drawn one above the other; this allows them to be compared more easily. In this case, there is *some* evidence of an interaction between the two species. This example is in fact to investigate the effect of trampling on plant growth across a footpath.

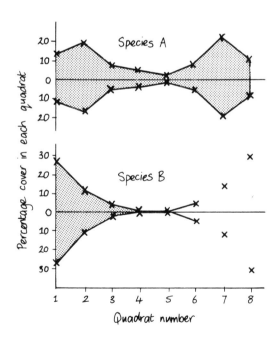

Fig 2.38
Example of a kite diagram

TABLES

Information from practical work in Biology is often recorded in a table. In fact, the table can be prepared *before* the experiment or exercise takes place, so that observations and measurements can be organised when they are actually made. To be able to do this, you must know what sort of information to expect from the practical. The table that is produced may be a suitable record of results. However, it may need to be 'cleaned up' and rearranged to help it provide a *clear* record.

The process of *tabulation* (putting information into a table) can be a very useful means of organising both measurements (data) and observations. Tables consist of two dimensions of information; vertical *columns* and horizontal *rows*. Construction of a table must firstly involve a decision about whether to arrange different sets of information in columns or rows. This may be decided simply by the amount of information. Long 'lists' of information are often placed in columns because there is likely to be more space available if the page is used vertically. (However, the page can of course be used horizontally).

By convention, results for the 'independent variable' (see under 'line graphs') are often written in the first column; all other columns relate to the 'dependent variable'. However, there is no particular need to follow this convention!

Related sets of information should all be arranged in the same way (e.g. 'photosynthesis rates' in the example below). To obtain high marks in the skill of tabulation, NEA candidates must be able to construct a table with at least two columns of data; also, the table must be fully labelled.

Figure 2.39 presents the results of an experiment to investigate the number of bubbles produced by an aquatic plant at different light intensities (see Ch. 4, Unit 16). The three repeat measurements ('trials') at each light intensity are shown. Notice that the results for each light intensity are kept separate. Note also that the table allows space for the mean ('average') values. Each set of figures is clearly labelled (with 'units') and the table itself has a title.

Fig 2.39
Example of a table: results of an experiment on the effect of light intensity on photosynthesis rate

Table showing the effect of light intensity on the rate of photosynthesis:

Distance between lamp and plant (cm)	rate of photosynthesis (no. bubbles/min).			
	Trial 1	Trial 2	Trial 3	mean value
30	2	8	5	5
25	6	10	11	9
20	17	13	18	16
15	20	28	31	26
10	35	34	35	35

(↓ increasing light intensity)

Tables should be kept as simple as possible and as clear as possible. Each table should ideally show one relationship, e.g. light intensity and the rate of photosynthesis. The experiment shown here could also include class results as well as results from the effect of temperature or CO_2 concentration on the rate of photosynthesis. Combining all these results in a single table, however, would be too confusing. Other examples of tables, some with examiner comments, appear in Chapter 4.

WRITTEN ACCOUNTS

'Scientific writing' has a quite distinctive style which students are likely to develop gradually during their courses in GCSE Biology (and in the other sciences, too). To gain higher marks in the coursework assessment, NISEC candidates should be able to demonstrate an ability to present both *verbal* and *written* reports on experiments or projects.

Written accounts are intended to be factual and objective ('detached'). For this reason, the style tends to be brief, 'clinical' and functional. It is written in the third person, past tense. For example, suppose an experiment involves adding 5 cm^3 of amylase solution to 10 cm^3 of starch suspension, then testing some of the mixture with iodine at regular intervals. The account of this might be written something like this:

1. 5 cm^3 of the amylase solution was added to a labelled test tube containing 10 cm^3 of 2% starch suspension. The tube containing this 'reaction mixture' was then placed in a water bath at 37°C. A stopclock was immediately started.
2. After 4 mins, 1 cm^3 of the mixture was placed into a clean test tube. 3 drops of iodine were added to this sample and any colour change noted. Immediately after the sampling, the 'reaction mixture' was immediately returned to the water bath.
3. The procedure described in (2) above was repeated at 2 min. intervals. The experiment was continued for a total of 7 mins.

Such an account (perhaps also accompanied with diagrams) is a fairly complete description of a particular experimental method. It does not include unnecessary details, but there is sufficient *information for another scientist to perform a very similar experiment*, using the above description as a set of instructions. This is, in fact, the main test of the quality of a piece of scientific writing.

CHAPTER 2 RECORDING AND COMMUNICATION

> **Your teacher will advise you which approach is most suitable for your cousework**

A common format for an experimental 'write-up' is as follows:

- **Title** – this should, in effect, summarise the *object* of the experiment.
- **Method** – this should describe the procedure that was used, with any precautions taken.
- **Results** – these should summarise all relevant observations and measurements, and also calculations (see Unit 6.1 below).
- **Conclusion** – this should involve an interpretation of the results, together with some assessment of the experiment (in terms of the objective stated in the title).

This sequence of method-results-conclusion is also the sequence by which the experiment is conducted. Each part is related directly to the others. However, they should not be confused (a common error unfortunately). The LEAG Examining Group states that 'the presentation of results, and any discussion following the experiment, should be clearly separated from details of the procedure'.

There are many variations on the method-results-conclusion format, however. For instance, some students will be asked to state *objectives* and/or *aims*, perhaps as part of an *introduction* or *summary* of the experiment, before the 'method'.

Another possibility is for a list of *apparatus and materials* to be included. These are often included in teachers' worksheets (see Fig 1.5), so that students know what to collect in order to perform the experiment. In addition to a 'conclusion', students may be asked for *criticisms* of the experiment, i.e. an evaluation of the method, perhaps also with suggestions for improvements and further work.

The general 'rule' for writing up is to keep it as brief as possible, and to use diagrams if appropriate. One intermediate solution is to use *annotated diagrams;* these are diagrams accompanied by brief 'notes' rather than just name-labels.

5.3 How are marks awarded?

Fig 2.40 shows how the skill is assessed.

Fig 2.40
'Recording and communicating': skill differentiation

ard:	LEAG	MEG	NEA	SEG	WJEC	IGCSE	NISEC
High	Drawing large, scale correctly given, correct relative proportions or correct numbers of distinctive features. Drawing neat, lines clear, distinct. Chosen features accurately described in a suitable form. Record of observations complete, clear presentation.	Record experimental data accurately and clearly and in the most appropriate form. Graphs, when plotted are correct in every respect.	**Skills available:** Skill 15 Using diagrams Skill 16 Completing prepared tables Skill 17 Devising an appropriate table to represent results including units. Skill 18 Plotting a graph on given axes which include the scale Skill 19 Constructing a graph, including choice of scale, labelling axes, units	Record results in an orderly manner	Make a clear accurate drawing showing all features in correct proportion.	Record experimental data accurately and clearly and in the most appropriate form. Graphs, when plotted are correct in every respect.	Write a concise report in prose of a biological experiment, project or event. Select the data to be recorded and decide on the most appropriate method of presentation. Process data by construction of charts and graphs
Mid	Drawing large, scale approximately correct. Proportions or numbers of features approximately correct. Drawing generally tidy. Chosen features described reasonably well. Most results or observations recorded. Overall presentation clear. Need prompting to adopt appropriate method.	Record information systematically and effectively given a simple format or guidelines.		Record results in slight disorder	Make a clear drawing showing most features; some drawn to the correct proportion	Record information systematically and effectively given a simple format or guidelines.	Give a concise verbal report of a biological experiment, procedure or project, and write a report of a biological experiment, project or event to include the points of importance. Record information in tabular form by means of words, symbols, figures and percentages without assistance.
Low	Drawing small, scale poorly recorded. Proportions or numbers of features bear some resemblance to the specimen. Drawing completed but lines sketchy. Suitable features chosen with difficulty.	Record observations and results accurately if given a tightly prescribed format.		Record results in considerable disorder	Draw only some features, with proportions bearing some resemblance to the specimen.	Record observations and results accurately if given a tightly prescribed format.	Describe a biological procedure or experiment verbally. Record information in tabular form by means of words and figures when given headings or categories.

5.4 Examples of practical work involving this skill

There follows a limited selection of the numerous possibilities for work involving recording and communication. All of these examples have been suggested by the GCSE Examining Groups, but it should be remembered that there are many other appropriate examples not included here.

- Production of accurate drawings to show differences in structure in different flower species.
- Prepare a table to show loss in mass of a potted plant over a period of time.
- Construct a table to show the results of a choice chamber experiment.
- Suitably present the results of a respirometer experiment, involving woodlice.
- Construct food chains or food webs from ecological sampling.
- Produce tables (and possibly graphs) of results showing changes in mass of potato tissue, due to osmosis.
- Plot a graph to show the relationship between light intensity and the number of bubbles produced by a pondweed.
- Produce a histogram to show continuous variation in privet leaves.
- Record, in a suitable way, the results of an experiment on reaction (using a falling metre rule).
- Make a drawing to show the results of an experiment on phototropic responses in a seedling.
- Draw a diagram of the apparatus used to demonstrate the production of CO_2 by a small mammal.
- Make a drawing of the dissected parts of a flower.
- Record the results of an experiment to show the conditions needed for germination.
- Record the movement of water in a capillary-tube potometer.
- Make a drawing of an unfamiliar specimen collected from a habitat.

UNIT 6   INTERPRETING INFORMATION

Interpretation is an important and often difficult skill to develop. It involves *processing information* (i.e. by calculations), *extracting information, recognising patterns* and *making deductions*. Each of these main sub-skills includes several other important abilities which may need practice.

Interpreting information is much easier if the *purpose* of the experiment is understood. For this reason, it is helpful if the student has also developed skills in experimental design (Chapter 3). The information being interpreted may arise directly from the results of an experiment that has been conducted. Alternatively, the student may be asked to interpret information provided by the teacher. However, some Examining Groups require that *both* recording and interpreting of information should be based on the *same experiment*. If a student fails to record information in an adequate form, this should not prevent him/her from demonstrating maximum ability in interpretation. For this reason, the teacher may assist at the recording stage. The teacher may even process information from an experiment into a suitable form.

Fig 2.41
'Interpreting information': skill summary

LEAG	MEG	NEA	SEG	WJEC	IGCSE	NISEC
Give suitable conclusions. Explain the results and show an understanding of their significance.	Make conclusions and logical inferences from data available.	Extract information accurately. Perform appropriate calculations. Recognise variability in measurements. Formulate a generalisation consistent with data.	(Skill not assessed in this form)	(Skill not assessed in this form)	Make conclusions and logical inferences from data available.	Extract information from data supplied. Draw correct deductions from results. *(Identify sources of error, criticise procedures, make suggestions for further relevant investigations).

*(more able candidates only).

6.1 Processing information: calculations

Students in GCSE Biology may be asked to perform some simple calculations, both in the written examination and in coursework. Practical work often involves measurements, and this will produce collections of values, or *data*. The data obtained directly from an experiment is sometimes called *raw data*. Raw data may need to be processed (not 'cooked'!) by a series of relatively simple calculations. In fact, such 'number-crunching' typically occupies a fair amount of time in experimental Biology. The use of calculators is permitted by Examining Groups for calculations in GCSE Biology (written examinations and coursework). There are two main reasons why these calculations may be necessary:

▶ The raw data may be difficult to interpret unless it is transformed into something more understandable.
▶ The raw data may consist of repeated measurements of the same situation taken, for example, to allow for variation. This data may need to be 'summarised'.

Examining Groups assume that GCSE Biology candidates will have had some mathematical background. The types of calculation which are likely to be most useful in Biology at this level are: *fractions, decimals, percentages, ratios* and *means* (averages) and *dimensions* (i.e. length, area and volume).

It is important to remember that calculations in Biology coursework are a means to an end. In other words, calculations are part of the overall process of interpretation. Examples of such calculations appear throughout Chapter 4.

6.2 Extracting information

Information is more easily *extracted* if it is in an accessible form. This means that the way in which the information is recorded and communicated (Unit 5) may determine how information is extracted.

'Extracting' means identifying and isolating key pieces of information from amongst other material. NEA candidates are asked to select particular pieces of information from at least twenty items, for example in a table, in graphs or pie charts. Students may be directed towards specific information by a series of *questions*. However, some Examining Groups regard the provision of questions by the teacher as equivalent to help being given. In this case, less marks will be available for this particular assessment.

Asking questions is, however, a useful method of extracting information from an array of results. Questions do not necessarily need to be supplied by the teacher. Students may be given the responsibility for formulating their own questions about the information. Often it is simply the answers which emerge (e.g. in the written account of an experiment), but such answers are provoked by a questioning approach towards the information. There are some fairly general questions which apply to most experiments, including:

▶ 'What was the original purpose of the experiment?'
▶ 'Has the purpose of the experiment been achieved?'
▶ 'Has the original assumption been proved?'
▶ 'What trends are apparent in the data?'
▶ 'What were the limitations of the experiment?'
▶ 'Could the experiment be easily improved?'

Each individual piece of practical work should also generate very specific questions which relate more precisely to the method and materials involved.

Some teachers will encourage class, group or individual *discussion* of results, so that students think carefully about the implications of their results, or results supplied by the teacher. Students, individually or in small groups, may be asked to explain their results to the class as a whole. In this way, such students may discover how well (or otherwise!) they have interpreted the information.

Some examples of extracting information are given below, using actual coursework data:

Example 1: Information in a table

The data in the table (Fig 2.42) show the effect of temperature on the rate of bubble production (= respiration rate) in yeast.

Fig 2.42
Table of data: the effect of temperature on respiration rate of yeast

time mins	1st	2nd	3rd	4th	5th	6th	7th	8th	9th	10th
no. of bubbles per min. (30°C)	0	0	1	2	2	3	4	5	6	5
no. of bubbles per min. (40°C)	10	9	9	8	6	6	6	5	4	5

The teacher asked the following questions on this data:
a) Why did it take some time for the bubbles to appear?
b) Why did the rate of bubble production slowly increase at 30°C?
c) Why was the rate of bubble production initially faster at 40°C compared with 30°C?
d) Why did the rate of bubble production start to decrease towards the end of the experiment at 40°C?

> You may need to quickly revise a particular piece of theory in order to interpret results in an appropriate way

To answer these questions satisfactorily, you would need to show an understanding of the processes involved. In this experiment, it is really *enzyme activity* (incidentally, 'enzyme' orignally meant 'in yeast'!) which is being investigated. Once this is realised, you can refer to the effect of temperature and also the concentration of substrate and product on enzyme activity. The substrate is broken down by enzymes to release energy. The substrate is used up more quickly at 40°C because enzyme activity is more rapid at that temperature. Also, a waste product (alcohol) is produced in yeast respiration; this is toxic and, when it accumulates, inhibits enzyme activity. You could suggest whether it is the depletion of substrate or the accumulation of toxic product which is having the most significant effect in this experiment.

You should, if possible, *refer directly to the data*. This is generally a good idea for all data interpretation. In this example, it could be pointed out for instance that the rate after ten minutes is the same (i.e. 5 bubbles per minute) at both temperatures. Mean values can be calculated; these are 2.8 for 30°C and 7.1 for 40°C; the mean rate is about 2.5 times more at the higher temperature. The 'conclusions' presented here are by no means complete, and are simply to outline some of the processes involved in extracting and using data.

Students may even have an opportunity to draw a *sketch graph* to illustrate their account, or simply during the rough planning stage in preparing a conclusion. These graphs break all the normal rules (i.e. they don't need to have titles or numbered scales) but they can be very effective! (Fig 2.43). Note that this sort of graph should only be used in interpreting data, not in recording and communicating results.

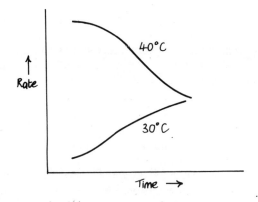

Fig 2.43
Sketch graph to illustrate an interpretation [based on the data in Fig 2.42]

Example 2: Information in a graph

The frequency histogram in Figure 2.44 shows the distribution of hand span in a class of twenty-three pupils of approximately the same age. The relative number of males and females was also noted. (The method for organising data of this type is described in Unit 5, pp 28–29).

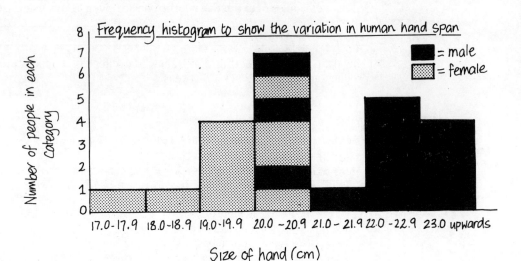

Fig 2.44
Frequency histogram: distribution of hand span in a group of male and female pupils

The following questions were asked about this survey. It is for the student to decide whether to extract the information from the histogram, or from the frequency table or the raw data (not shown here).

▸ For *all* the data, what was the range of values (i.e. minimum and maximum) and what was the mean value?
▸ For the *male* and *female* groups, what were the range and the mean value in each group?
▸ For *all* the data, what was the modal size group (i.e. the size group with the highest frequency)?
▸ Comment on the distribution of values that you have identified.

The information extracted from the graph and from the data itself reveals some fairly strong trends (see below) which can be referred to. For instance, there is *variation* in the values; the range for all the data 17.6 – 23.8 is quite large; the 'size' of the range is the difference between these numbers, i.e. 6.2 cm. This could be due to 'growth differences' between individuals, or to differences in measurement techniques (all students measured their own hands). Male/female differences are very striking in the graph, with only a limited band of overlap. Some possible explanations for this could be suggested. More general comments on the size of the sample might also be made.

Example 3: Information from a line graph

Line graphs are not simply a means of displaying data (Unit 5); they can also be used to extract information. There are several ways of doing this:

▸ **Trends** Graphs can reveal overall trends in a set of data. The shape, slope and position of a curve relative to the axes (scales) can be very informative. It should be remembered that any graph is simply a way of showing the relationship between two variables, represented by the two axes. Graphs can also be very useful in comparing different sets of data, plotted as separate lines on the same axes (Fig 2.45).

Fig 2.45
Line graph: relative pulse rate of boys and girls

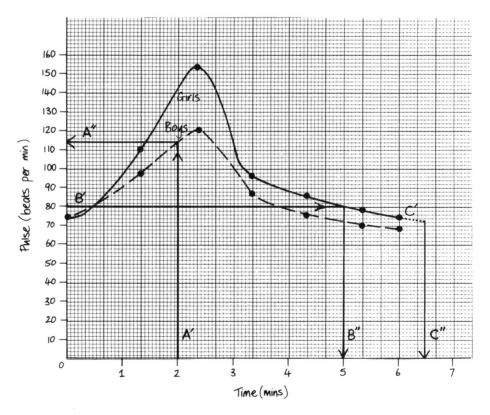

▸ **Interpolation** Interpolation means extracting information from a graph *between* points. The points on a graph were obtained by actual measurements during an experiment. These points are connected by a line (straight or curved). It assumes that if any 'intermediate' measurements had also been made, they would fall on this line. The graph can therefore be used to *estimate* the probable value between any two points.

For example, in Figure 2.45 the mean pulse rate for boys can be estimated for two minutes by placing a ruler vertically at the 2 min position on the x-axis, indicated by A'; a line could even be drawn onto the graph. Where the vertical line meets the curve for boys pulse rate, a horizontal line can now be drawn; line A". Where this meets the y-axis gives the corresponding pulse rate; it is about 114 beats per minute. Interpolation can of course operate the other way round. For example the time at which the average pulse rate during the *recovery* stage (i.e. curve 'descending') in girls reaches 80 beats per minute (line B') can be found. This occurs at about 5 min (line B").

- **Extrapolation** Extrapolation means extracting information from a graph *beyond* the points. In other words, 'off the end' of the graph. For instance, the region marked with a dotted line on the graph for girls represents probable values which might be obtained if measurements had continued. Extrapolation allows *predictions* to be made. In the example shown in Figure 2.45, the dotted extrapolation (line C') shows that pulse rate in girls might reach the 'normal' resting level of 72 beats per min at about 6.5 minutes (line C"). All extrapolation should be done with caution, however, as there may be unexpected departures from the observed pattern of the graph.

In general, line graphs become more reliable as the number of points used to plot them increases. The minimum number of points (coordinates) used to construct a straight line graph is three. However, these points may not necessarily be truly representative (especially if they are close together). The use of additional points is therefore recommended. For curved (non-linear) graphs this is certainly necessary. More frequent measurements may need to be made during experiments to allow for this.

6.3 Recognising patterns

Some reference has already been made to the recognition of patterns, as part of the process of extracting information. There are two main approaches to this:

'INDUCTIVE THINKING'

This means making a general interpretation from particular pieces of evidence. For example, suppose a student conducts several experiments with different enzymes; in all cases the rate of reaction was found to be more rapid at higher temperatures. A general interpretation from these separate pieces of evidence might be that *enzyme activity increases with increasing temperature*.

'DEDUCTIVE THINKING'

This means seeking particular pieces of evidence to support a general assumption. For example, suppose a student *assumes* that, in general, enzyme activity increases with increasing temperature. The student may then seek evidence to support (or refute) this assumption in the results of an experiment, using a particular enzyme.

Both these approaches are quite valid. Indeed in Biology practical work it is normal to use a *combination* of the two, though it is fair to say that the main approach tends to be *inductive*. The actual choice of approach may in practice depend on:

- how clearly the teacher had defined the purpose of the experiment (intentionally or otherwise) or,
- how well the student has understood the purpose of the experiment; this may in turn depend on their familiarity with the *theory* involved.

An *inductive approach* might be used if the teacher has set an 'open-ended' practical, where students discover patterns as a result of conducting the experiment and analysing the results, i.e. 'An experiment to investigate the relationship between . . . '.
A *deductive approach* might be used if the teacher has provided a firm objective, which the students confirm, i.e. 'An experiment to prove that . . . '.

Students should, of course, resist the temptation to discover non-existent trends or pieces of evidence in their results which confirm a previous assumption. It may be that the results don't 'turn out' as expected and that the experiment 'hasn't worked'. If so, the student should interpret the results accordingly. This can be quite difficult, because it may mean abandoning a simple, neat explanation in favour of an uncertain and 'messy' conclusion! This problem will be covered in more detail in the next chapter.

DESCRIBING RELATIONSHIPS

Many experiments in Biology are designed to show the effects of one variable on another. There are three main types of possible relationship (or 'correlation'). They are shown in Fig 2.46):

- **Positive relationship**

As one variable increases, so does the other; this is also called a 'direct' relationship. Example: the growth of barley seedlings (Ch. 4, Unit 9).

> 66 **Don't worry if experiments give unexpected results – marks are given for interpreting what *actually* happens, not for what 'should' happen** 99

▶ **Negative relationship**
As one variable increases, the other decreases; this is also called an 'inverse' relationship. Example: the strength of glucose solution and the length of potato cylinders (Ch. 4, Unit 4).
▶ **Neutral relationship**
As one variable increases or decreases, there is no obvious corresponding change in the other. This sort of relationship is admittedly unusual at GCSE level, because practicals are often designed to produce 'strong' results, which perhaps gives more scope for student interpretation.

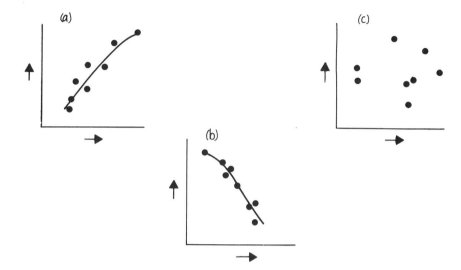

Fig 2.46
Three types of relationship in line graphs: (a) positive, (b) negative relationships, (c) neutral relationship

Both positive and negative relationships vary in 'amount', reflected by the slope of the line. The lines in Figure 2.46 (a) and (b) are fairly straight; a change in one variable is accompanied by a uniform change in the other. This is known as a *linear* relationship. However, it is possible that a change in one variable is not always accompanied by a constant change in the other. The result is a curve, known as a *non-linear* relationship.

The terms used here are for convenience only. The student should describe relationships in the data using words that are familar and accurate. Direct reference to the data is always a good idea; this shows that the student is actively interpreting *actual results*, not re-writing the object or aims of the experiment. However, Examining Groups ask students to avoid simply re-stating original data.

RECOGNISING VARIABILITY

Variability in results is fairly common in Biology experiments. A significant source of this variability is likely to be in the organisms or tissues which are being investigated. For instance, microscopic examination of epidermis cells (Ch. 4, Unit 3) often reveals considerable variation in the size and shape of cells. Also these cells may respond quite differently to a concentrated external solution (Ch. 4, Unit 4).

However, students should be aware that variability in results can also result from experimental error (Unit 5.1, p. 24). The sources of this error may for instance be in the accuracy of measurements or in calculations. An important skill therefore is to identify 'unusual' or 'unexpected' results. These results may stand out because they seem out of place in relation to other results from the same experiment. This can occur between or within sets of data:

Between data sets

A whole set of results from one group may be strikingly different from those of other groups in the same class. In the example shown in Figure 2.47, differences between the results from Pupil A and Pupil B might be due to differences in the activity or concentration of their salivary amylase, or due to experimental technique. Variations can also occur between data sets when the experiment is repeated by the same group.

Within data sets

For the same set of measurements taken by one group, a particular value may not seem consistent with others in the same data set. In the example shown (Fig 2.47), 'anomalous' results in Pupil B's results may be due, for instance, to incomplete sampling or mixing. This may only become apparent when *trends* in the data are being identified. Graphs can be a very useful way of revealing experimental error (Fig 2.48).

Fig 2.47
Table of results containing experimental variability (data supplied by NEA and published with permission)

Temperature °C	Time for Starch to Disappear	
	Pupil A	Pupil B
0	Starch still present at 30 min	Starch still present at 30 min
10	8 min 30 s	10 min 0 s
20	4 min 0 s	2 min 30 s
30	1 min 30 s	3 min 0 s
40	0 min 30 s	1 min 30 s
50	0 min 30 s	Starch still present at 30 min
60	Starch still present at 30 min	Starch still present at 30 min

Fig 2.48
Graph of results, revealing experimental variability (based on results of Pupil B, Fig 2.47)

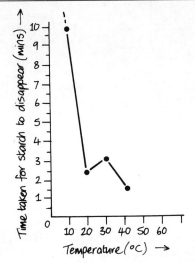

This skill relies, to a large extent, on a student's understanding of experimental objectives and procedures. The identification and explanation of variability or experimental error *can* be difficult. Examining Groups in fact recognise this as a 'high-level' skill; it is expected that only a relatively small proportion of students will show competence with this.

6.4 Making deductions

This skill involves an ability to draw valid conclusions from *evidence contained in the data* only. In other words, the conclusion should be *directly supported by the evidence*. This means:

▶ that all the data should either be referred to in a general or a specific way, and
▶ that related or relevant information which does not appear in the results should not be used to support the conclusion.

However, students may still demonstrate their practical and theoretical understanding of the experiment in the way that they interpret their results. For instance, pupil B in the example above (Fig 2.47) may be able to make this sort of conclusion:

"As the temperature increased from 0°C to 40°C there was a fairly steady decrease in the time it took for the starch to disappear. The time at 30°C seems a bit high in comparison, if the results for 20°C and 40°C are accurate. This may be because the sample was not properly mixed with the iodine solution.

The time taken for starch to disappear increases rapidly at 50°C and 60°C. This suggests that the disappearance of starch is probably due to the presence of an enzyme in the saliva. The changing rate of starch breakdown in this temperature range (0°C – 60°C) is characteristic of enzyme activity".

Many of the comments on interpretation refer to data derived from *measurements*. Conclusions may also be drawn from *observations*. Figure 2.49 shows a student's results of an experiment on the gain or loss of CO_2 by leaves. The experiment is based on pH-sensitive bicarbonate indicator. The indicator can be used to monitor CO_2 concentrations. Its colour is red at neutral pH ('normal' CO_2 levels). There is a colour change to yellow at acid pH (high CO_2) and purple at alkali pH (low CO_2).

The student has used this information and the actual colour changes observed to propose some conclusions (Fig 2.50). These conclusions relate both to CO_2 levels and, by inference,

CHAPTER 2 INTERPRETING INFORMATION

possible explanations for these. Note the use of a table to summarise the conclusions for this experiment.

This does not, however, represent a complete interpretation of the results. For instance, it is important to compare experimental tubes with their corresponding *controls* (see Chapter 3). Such a comparison could be given in a written account, accompanying the table shown here.

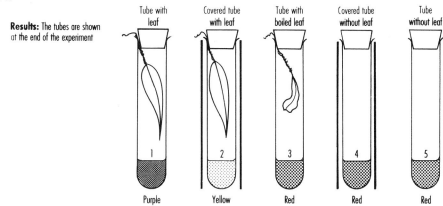

Fig 2.49
Results diagram: CO_2 uptake and loss by leaves

Fig 2.50
Summary table: CO_2 uptake and loss by leaves (based on the results shown in Fig 2.49)

Summary of conclusions

Tube number	Colour	CO_2 Level	Conclusion
1	Purple	low?	(Not very clear) CO_2 absorbed for photosynthesis
2	Yellow	High	CO_2 produced by respiration
3	Red	Medium	No effect: the leaf is dead so no respiration or photosynthesis occurs
4	Red	Medium	No effect ⎫ No leaf present to do
5	Red	Medium	No effect ⎭ anything

Fig 2.51
'Interpretation of information': skill differentiation

6.5 How are marks awarded?

Fig. 2.51 shows how this skill is assessed.

LEAG	MEG	NEA	SEG	WJEC	IGCSE	NISEC
State conclusions concisely. Draw clear understanding of inferences. Carefully explain results, including reference to controls. Comment usefully on limitations of methods or indications of the wider significance of results.	Make accurate conclusions and logical inferences from data available.	**Skills available:** Skill 20 Extracting information from tables and charts Skill 21 Extracting information from graphs Skill 22 Performance of appropriate calculations on experimental data Skill 23 Identifying anomalous results Skill 24 Identifying areas of experimental error and explaining any variation in results Skill 25 Recognising patterns Skill 26 Ability to draw valid conclusions from data	(Skill not assessed in this form)	(Skill not assessed in this form)	Make accurate conclusions and logical inferences from data available	Extract information from data supplied in a wide range of sources. Draw correct deductions: make reasoned judgements about alternatives; identify sources of error and make suggestions for improved procedures on further relevant investigations.
State conclusions simply. Provide some explanation of results. Provide a limited discussion of the significance of results in relation to methods used.	Interpret data to produce logical conclusions only with considerable assistance				Interpret data to produce logical conclusions only with considerable assistance	Extract information from data supplied in a variety of sources. By means of relevant questions, draw correct deductions from experimental work; use the information to criticise procedures and results.
Draw suitable conclusions with assistance. Attempt to explain results but show limited understanding of their significance	Given considerable assistance, recognise a limited number of relationships within the data available				Given considerable assistance, recognise a limited number of relationships within the data available.	Extract information from data supplied in a variety of sources. Draw correct deductions from results generated or supplied within a biological context and relevant to pupils own experience, by means of appropriate questions.

6.6 Examples of practical work involving this skill

The following examples, all suggested by Examining Groups, represent a small sample of the many suitable coursework practicals which can be used to assess the skill of interpretation of information.

- Calculate the mean (average) mass of a batch of germinated seedlings.
- Describe the relationship between light intensity and the rate of bubble release in a pondweed.
- Calculate the percentage of humus in a sample of soil.
- Determine the components of some common food from tables.
- Calculate the relative percentage of exhaled gases, determined by analysis using a gas burette.
- Describe the relationship between exercise and pulse rate.
- Calculate changing trends in human population, using a table of data.
- Describe the relationship between the rate of starch digestion and temperature.

UNIT 7 CARRYING OUT SAFE WORKING PROCEDURES

Students are expected to work safely at all times. However, this important skill may only be actually assessed once, or perhaps not at all, depending on the Examining Group (Fig 2.52). For most Groups, 'safety' is incorporated in the assessment process within 'Following instructions' and 'Handling apparatus and materials' or equivalent (Units 2 and 3). SEG is the only Examining Group to specify 'safety' as a separate skill, and the assessment of this skill is expected to occur towards the end of the Biology course.

Fig 2.52
'Carrying out safe working procedures': skill summary

LEAG	MEG	NEA	SEG	WJEC	IGCSE	NISEC
Work safely	Show due regard to safety	Pay attention to safety.	Carry out safe working procedures (after initial instruction to the class).	Assemble and/or use apparatus and/or materials safely.	Show due regard to safety.	Be aware of laboratory hazards. Use a range of scientific apparatus safely.

Fig 2.53
Carrying out safe working procedures: skill differentiation

Standard:	LEAG	MEG	NEA	SEG	WJEC	IGCSE	NISEC
High	Work safely.	Safety procedures constantly and systematically incorporated into work.	(Skill not differentiated in this form). Skill 11 includes: Attention must be paid to safety.	Work safely without reminders	Manipulate apparatus and/or materials safely without assistance.	Safety procedures constantly and systematically incorporated into work.	Be fully aware at all times of laboratory hazards. Use a wide range of general scientific apparatus safely.
Mid	Show awareness of safety precautions, but sometimes careless	Errors and dangers recognised and remedied when these are pointed out.		Work safely with rare reminders	Manipulate apparatus and/or materials safely with some assistance.	Errors and dangers recognised when these are pointed out.	Be aware of laboratory hazards. Use a wide range of general scientific apparatus safely
Low	Use safety precautions when reminded			Work safely with several reminders	Manipulate apparatus and/or materials correctly and safely only with considerable assistance.		Note all potential major hazards. Assemble apparatus safely. Use general science apparatus safely.

UNIT 8 USEFUL READING

The following books cover the practical aspects of Biology, mostly at the GCSE level. Those marked * are intended for more advanced students, or for teachers, but may still be very useful for students of GCSE Biology.

- *Biology Practicals* * Jack Hardie and John Tranter (Longman, 1987)
- *Biology for Life* * John Finagin and Neil Ingram (Nelson, 1988)
- *Experimental Work in Biology* (Combined Edition) D G Mackean (Murray, 1983)
- *Inquiry and Investigation in Biology* * A.B.A.L. (Cambridge University Press, 1983)
- *Investigations for GCSE Biology* P W Freeland (Hodder & Stoughton, 1987)
- *Life Story: Experiment Guide* F M Sullivan (Oliver & Boyd, 1986)
- *Practical Biology* * M K Sands and P E Bishop (Bell & Hyman, 1984)

CHAPTER 3

TACKLING STUDENT-ORGANISED COURSEWORK ASSESSMENTS: EXPERIMENTAL DESIGN AND PROBLEM-SOLVING

UNIT 1 — STUDENT-DIRECTED PRACTICAL WORK

GCSE has continued and intensified a general shift in emphasis towards more student-centred activities. GCSE students are expected to take more direct responsibility for their learning, and they are given credit for this. In GCSE Biology, you are encouraged to take an active part in planning, conducting and evaluating your own practical work. This is assessed by all Examining Groups as a separate skill, often called *Experimental design and problem solving* (see Fig 1.4):

Examining Groups are *not* expecting students to conduct large and ambitious projects; the skills being covered in this chapter can be effectively demonstrated in '*small-scale investigations*'. Students should remember that, depending on the Examining Group, 'experimental design' is not likely to be allocated much more than about 5 per cent of the total mark allocation for GCSE Biology. These *are* certainly marks worth obtaining, but you should not spend *too* long on your investigations; time and energy will be needed for other aspects of GCSE Biology (and for other subjects, too!)

Examining Groups permit groups of students to devise a plan for investigating a single problem but, as with all coursework, each student is assessed individually for each skill, e.g. in planning and evaluating the investigation.

Although 'experimental design' is assessed separately from other skills, this does not mean that it is developed in isolation from them. This important and challenging skill is developed gradually throughout the course, as students learn the techniques required during teacher-directed practical work (Chapter 2). Students are normally expected to make any decisions whilst doing experiments and exercises organised by the teacher. For instance, choosing which method to use (there may be several alternatives), deciding on suitable ways to make observations and measurements, selecting the most appropriate way to record the results and so on.

Teachers often plan their practical work in a fairly flexible way. This gives students the opportunity to pursue lines of enquiry which may develop out of the original practical. You will gradually learn that there are frequently no 'right' or 'wrong' answers in science. As part of your scientific training, you should be ready to respond to the implications of your own observations and deductions. This is why teachers often encourage students to *evaluate* a particular experiment. You may then be able to *suggest improvements* and *further work*. There may be practical difficulties why an experiment cannot be continued in this way, such as a lack of time or resources. However, it is still very useful for students at least to think about the ways in which an investigation could be followed up.

Fig 3.1
'Experimental design and problem solving': skill summary

LEAG	MEG	NEA	SEG	WJEC	IGCSE	NISEC
Formulate a hypothesis and plan. Choose suitable methods. Show awareness of need for validity, eg. repetition, controls. Show awareness (and understanding) of limitations in method and/or hypothesis. Suggest improvements (and further work).	Identify the problem and plan the investigation. Choose techniques, apparatus and materials. Organise and conduct the investigation. Evaluate methods and make positive suggestions for improvements.	Criticise experimental design, including the use of controls. Formulate hypotheses. Devise an experiment to test a hypothesis.	Design an experiment to test a hypothesis.	Design and conduct an experiment to test a formulated or given hypothesis.	Identify the problem and plan the investigation. Choose techniques, apparatus and materials. Organise and conduct the investigation. Evaluate methods and make positive suggestions for improvements.	(Propose a relevant hypothesis). Decide what to measure or observe. Select appropriate apparatus. Control one (or two) variables. Evaluate the method and result (and change the methods or design accordingly).

CHAPTER 3 IDENTIFYING A PROBLEM AND PLANNING AN INVESTIGATION

This chapter covers the particular skills required for experimental design and problem solving. This means that you are given the opportunity to show an 'ability to design and carry out investigations based on a given problem' (MEG/IGCSE). A complete chapter is used for this skill because, by definition, it involves much more student responsibility than the other, teacher-organised skills being assessed (Chapter 2).

UNIT 2 IDENTIFYING A PROBLEM AND PLANNING AN INVESTIGATION

2.1 Identifying a problem

Depending on the Examining Group and on the abilities of the pupils involved, teachers may give a fair amount of guidance in the early stages of planning an investigation. A very common starting point is for the teacher to suggest a single topic or group of topics (often related) which are likely to be suitable for investigation. The topic may arise naturally out of other work (including coursework).

Alternatively, the teacher may present ideas specifically for the assessment of this skill, for instance towards the end of the course. A further possibility, which can be both challenging and rewarding, is for students to generate their own problems; some suggestions on how to tackle this are provided in Unit 6.

'Identifying a problem' essentially means locating the possibilities for experimental investigation within a piece of biological information. For instance, this information may be presented to students in the form of a statement. There are several such statements listed at the end of this chapter. Here is one example, which will be developed throughout this chapter:

▶ "Nettle plants grow differently in the shade compared with being in the sun".

There are several problems here. The main problem seems to involve the effect of different light intensities on growth. The actual *process* of growth is relatively slow (even in nettle plants!), so you might decide to investigate the *outcome* of growth. There are different ways of measuring growth, and it may be that different students (or groups) decide to measure different aspects. There are also *practical problems* to be overcome. For instance where, what and how to sample growth in groups of nettle plants (Fig 3.2).

Fig 3.2
Measuring leaf length of shaded nettle plants

Students will be given additional credit if they can identify a problem without assistance from the teacher. There will also be opportunities to formulate a hypothesis (Unit 4.1) for this problem, to select or even construct apparatus. The investigation may also involve the assessment of other skills, such as in conducting the actual investigation and recording and interpreting the results.

2.2 Planning an investigation

Having identified at least one problem you must next plan the actual practical work. A common difficulty is that students tend to make very ambitious plans. It may not be possible for such plans to be carried out because of a lack of time or resources.

Note that Biology experiments often depend on the availability of 'biological material'. For instance, experiments involving flowers and fruits often can only take place at certain times of the year. An investigation involving germination or phototropism in plants or development in insects or tadpoles would need to be planned well in advance to allow for the growth period. Another reason why plans are sometimes too ambitious is that students can underestimate the skills needed to conduct large or complex investigations. It is perhaps easier, therefore, to *begin with a fairly modest, open-ended idea*. This can gradually be developed, as opportunities permit.

> **Investigations need to be planned with care if they are to be successful**

FORMING THE PLAN

Here is one possible sequence in which a plan could be developed:

1   Make a rough, general plan

The plan need not be very detailed, but could be used to *define* the area of interest, and to suggest the possible method and materials to be used. This plan can be discussed with others, including other students, the teacher and parents too! For example, you could decide to try measuring the heights of different nettle plants.

2   Plan a simple 'pilot' experiment or exercise

This can be used to check that the basic idea behind the investigation is workable and that materials are available. For example, taking a brief look at nettle plants growing in shade compared to those growing in sun.

3   Formulate a hypothesis (see Unit 4)

This could be the starting point for students who already have a fairly clear idea about their intended investigation. However, the hypothesis may need to be modified later.

4   Choose your apparatus

Credit is often given for the *appropriate choice of apparatus*. This may involve selecting the most suitable apparatus from a collection of apparatus which the teacher has assembled. You may be expected to prepare a list of the apparatus that you intend to use for your investigation. You may also be asked to discuss your requirements with the laboratory technician.

You should obviously avoid asking for apparatus that is not likely to be available or easily obtainable. An investigation can be a good opportunity to learn how to use new apparatus, but care should be taken that this does not become too time-consuming.

5   Plan the details of the sequence of the experiment

NEA candidates are given credit for this. Perhaps the best way of presenting the proposed sequence is as a list. For example:
1   Locate sunny/shady sites where nettles are growing.
2   Randomly select 20 plants at each site.
3   Measure and record the heights (from ground level to upper top of plant in each case).
4   Calculate mean height values for each site.

UNIT 3   EXPERIMENTAL DESIGN AND THE USE OF CONTROLS

3.1 Experimental design

At this stage it is perhaps worth attempting to make a distinction between an *experiment* and an *exercise* in practical Biology. Both involve making observations and measurements. The main difference is that an experiment involves a *comparison* of two or more situations, an exercise does not. This distinction may seem unimportant, but students can waste a lot of time trying to 'design' an exercise rather than an experiment! Figure 3.3 shows a few exercises and some very similar experiments to compare them with.

CHAPTER 3 EXPERIMENTAL DESIGN AND THE USE OF CONTROLS

The distinction between an 'exercise' and an 'experiment' is not always easy. For example, determining the energy content of a peanut (Ch. 4, Unit 17) would be generally accepted as an 'experiment' although only one situation is involved. However, this would be rather limited as a subject for experimental design. A comparison of the relative energy content of two or more foods would give much more scope!

Experiments involve changing conditions or situations. These are called *variables*. Any individual experiments should have *no more than two variables under investigation*. All other conditions which could vary must be *fixed* during the experiment (see below). Each of two *experimental variables* in any experiment changes in a different way and each can be described differently:

▶ **Independent variable:** this is changed by the *experimenter* (see Column 3, Fig 3.3)
▶ **Dependent variable:** this is changed by the *experiment* (see Column 4, Fig 3.3)

Fig 3.3
Comparison of some exercises and experiments.
Experimental variables are shown

Note that these terms have already been referred to in connection with graphs (Unit 5.2, p. 28).

1 Exercise	2 Experiment	3 Experimental variables — independent variable	4 Experimental variables — dependent variable	5 Examples of *fixed* variables.
[A] determine the pH of a soil sample	[A] determine the pH of three different soils, and compare them	different soils	pH values	amount of soil used, test method
[B] staining a stem section for microscopic examination*	[B] comparing the effect of two different stains on the same type of tissue	different stains	appearance of stained section	type of tissue used, preparation of mount
[C] dissecting and drawing a half-flower*	[C] dissecting and drawing flowers of the same type of plant at different stages of development	time that the drawing is made	appearance of flower	same species of flower used
[D] measuring the mass of a potted plant once	[D] measuring the mass of a potted plant at regular intervals for 20 mins	time that plant is weighed	mass of plant each time	environmental conditions. Loss of water from soil
[E] estimation of vitamin C in a piece of potato	[E] estimation of vitamin C in two pieces of potato, one fresh, the other boiled	whether potato is fresh or boiled	vitamin C content	relative amount of potato used. Temp. at time of testing
[F] testing milk for microbes	[F] testing for microbes in milk stored at different temperatures	temperature of storage	amount of microbe growth	source of milk, sampling and testing technique
[G] determining the energy value of a peanut	[G] comparing the relative energy content of different foods	type of food	energy content	relative amount of food used
[H] number of duckweed in a sample from a stream	[H] relative number of duckweed in samples from polluted/unpolluted streams	whether stream is polluted or unpolluted	relative number of duckweed	other environmental conditions of stream. Sampling methods

*(these exercises would not normally be conducted as experiments)

> **This aspect of experimental design is both complex and important! You may need to re-read this section carefully**

So, if the student is confident that a planned investigation *is* in fact an experiment, the experimental design can begin with the identification of the independent and dependent variables. At least one (the independent) of these two variables is expected to change during the experiment. The dependent (responding) variable will probably change too, but this is by no means certain. For instance, if the pH of several soils is tested, the results might show that they all have the same pH!

Whilst the experimental variables are allowed to change during the experiment, the other variables, called *fixed variables* should be kept as constant as possible. Such experiments are sometimes known as *controlled experiments*, and result from careful experimental design. However, experiments can only be designed within sensible limits, and it can be very difficult to keep all fixed variables constant throughout the experiment. This is particularly true for investigations which are not conducted in the laboratory (see below).

CONTROLLED EXPERIMENTS

Perhaps the most convincing case for careful experimental design is an example of an experiment which was *not* properly controlled. The following experiment was designed to show the effect of storage at higher temperatures on microbe growth in milk (Fig 3.3). A student decided to determine the relative microbe content of two samples of milk, stored at 4°C and 30°C. The milk can be tested for microbe activity by using resazurin dye; this dye is decolourised (from purple) by lactic acid, produced by any bacteria that are present. Some of the fixed variables for this experiment are given in Figure 3.3 (Column 5). Here is a brief summary of the student's experiment:

1 Description of milk used
Milk A was pasteurised, bought 4 days before the experiment and stored in a fridge (4°C).
Milk B was 'Longlife', bought 2 weeks before the experiment and kept at room temperature (20°C).

2 Conditions during the experiment
200 cm^3 of Milk A was stored in an open beaker *at 30°C in an incubator** for 48 hours.
500 cm^3 of Milk B was stored in its unopened carton *at 4°C in a fridge** for 3 days.

3 Method of testing
2 cm^3 of resazurin dye was added to 10 cm^3 of Milk A. *The colour of the dye was then noted #*.
5 cm^3 of resazurin dye was added to 5 cm^3 of Milk B. *The colour of the dye was then noted #*.

Interpreting the results of this experiment would be a biologist's nightmare! The independent variable (*) and the dependent variable (#) are shown. All other variables *should* be fixed. There are at least fourteen 'unfixed' variables in this brief description! (Can you spot them?) Each of these 'unfixed' variables is effectively a separate experiment in its own right.

UNCONTROLLED EXPERIMENTS

It is important, in designing experiments, to *work within reasonable limits*. These limits include the time and resources available, and your ability level. In other words, it may sometimes be more important to *keep the investigation simple and manageable* than to achieve perfect experimental design!

For example, in the experiment to compare the growth of nettle plants in sunny and shady conditions, the environmental difference being compared is obviously *light intensity*. However, there are also likely to be differences in *temperature, humidity* and *wind speed* between the two sites. There may also be differences in *water availability* and in the *soils*. Nevertheless, it is still possible to conduct a very worthwhile investigation, despite these difficulties.

Some allowances can be made for the possible effects of the additonal variables in the above experiment:

▶ Sampling between sites which are closely situated to each other and where additional variables are as similar as possible.
▶ Sampling at intervals along a transect ('sampling line') to see if there is a gradual change in any of these factors.
▶ Making brief measurements of additional factors; any variation between the sites can be taken account of during the interpretation or evaluation of the investigation.

A more 'thorough' approach might eliminate many of these variables, but would probably be too complex for GCSE. For example, nettles could be grown from an early stage in the laboratory, where the only environmental variable was light intensity. At GCSE then, some variables which are not intended to be part of the investigations may be too difficult to control, and you may need to *ignore* them. You should, however, demonstrate an *awareness* of unfixed variables, for instance in the written account of the investigations.

> **A 'perfect' experimental design is often difficult and time-consuming to achieve. It is a good idea to perform a 'realistic' experiment and comment on its limitations**

To summarise so far, experimental design and planning involves:

▶ Identifying the problem to be investigated.
▶ Checking that the investigation is an experiment rather than simply an exercise.
▶ Deciding on the experimental variables, i.e. what will be changed by the experimenter, and what result will be measured or observed.
▶ Identifying and, if possible, fixing any other variables.

3.2 Controls

What are controls?

Controls are duplicate experimental situations, identical in every respect except for the variable being investigated, which is kept constant. *Any difference in the outcome of the control and the 'experiment' must be due to the variable being investigated*. In other words, the independent variable (see above) is the only difference in experimental design between the 'experiment' and its control. If there is a change in the dependent ('responding') variable in the experiment but not in the control, then it is fairly safe to assume that one variable affects the other.

Are controls always needed?

No. In experiments which involve *comparisons* between different situations involving the independent variable, it is not always necessary to use a control. In these experiments all fixed variables are kept constant (see above); in a sense, each 'situation' is effectively a control for all the others. For example, in an experiment to study the effect of different temperatures on enzyme activity (Ch. 4, Unit 2) the only differences between each experimental condition are in the independent variable (temperature). Another example of an experiment involving comparisons but no control is an investigation of the effect of different light intensities on the rate of photosynthesis (Ch. 4, Unit 16).

How can I be sure which, if any, control to use?

Some students find controls difficult to understand. It is fair to say that the concept of 'controls' *is* quite demanding! The skill of identifying appropriate controls is very useful, and is normally developed gradually throughout the Biology course. Any student who is unsure about how to design an experiment with suitable controls should look at some examples in this and other GCSE Biology books (see the booklists on p. 7 and 43). Also, it is a good idea to discuss controls and experimental design with the teacher.

Here are a few more examples of the use of controls in GCSE Biology:

▶ **Example 1: To investigate the effect of higher temperatures on microbe growth in milk**

The experimental situation is the milk stored at the higher temperature. The control is the milk stored at 4°C; this is the 'standard' condition with which the experimental result can be compared. The milk samples should be from the same original source and they should both be tested in the same way for microbe activity.

If the purpose of the experiment was to study the effects of storage at different temperatures, then each experimental situation would effectively be the control of the other. However, the fixed variable must be kept constant.

▶ **Example 2: To investigate the relative effect of shade and sun on the growth of nettle leaves**

This is another example of an investigation without an obvious control because it is a comparison between two equivalent experiments. Each experimental situation is the 'control' of the other. If the investigation is 'to study the effect of shade on the growth of nettle leaves' (note the difference from the title above!) the control would be growth of nettle leaves in the sun.

It is probably not a good idea to refer to 'normal' conditions in this sort of situation. Some species of plant 'prefer' sunny conditions. Other species 'prefer' (i.e. are better adapted for) shady conditions. Still other plants are not particularly specialised for either condition compared with the other.

▶ **Example 3: To show that chlorophyll is needed for the production of starch**

'Variegated' leaves do not have chlorphyll distributed throughout their tissues. Such leaves can be tested for starch, using iodine. The position of starch is shown by a characteristic blue-black colour. This is found to correspond exactly with the distribution of the chlorophyll in the leaf, before testing (Fig 3.4). In this experiment, there is a 'built-in' control. These are the areas of the leaf where there was originally no chlorophyll, and which do not correspond with the presence of starch.

Fig 3.4
Experiment to show that chlorophyll is needed for starch formation.
(a) Before testing; (b) After testing

UNIT 4 FORMULATING AND TESTING A HYPOTHESIS

4.1 Forming a hypothesis

A hypothesis is an assumption, an 'educated guess', based on observations. A hypothesis can be *tested* by further observations. These may allow the scientist to *confirm* ('accept') or *contradict* ('reject') the hypothesis. If necessary, the hypothesis can be *modified* ('adapted') as a result of further observations. A biologist may therefore use a succession of observations and hypotheses (Fig 3.5). Each stage involves a slightly different approach. Observations require *critical* and *analytical* thinking. Hypotheses result from *imaginative* thinking. The process involves making observations, from which a hypothesis can be formed and tested. Further observations may then result in a new, modified, hypothesis being formed.

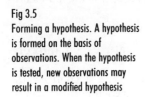

❝ Forming a hypothesis can be difficult and may need practice ❞

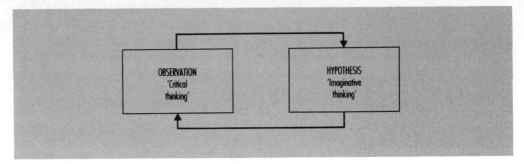

Fig 3.5
Forming a hypothesis. A hypothesis is formed on the basis of observations. When the hypothesis is tested, new observations may result in a modified hypothesis

Students may be presented with a problem (see Unit 2.1, p. 45) by the teacher and asked to suggest a suitable hypothesis. Another possibility is that a problem may emerge during laboratory or fieldwork, with students then asked to provide an appropriate hypothesis. There are several important things to remember about any hypothesis:

▶ It should make reasonable biological sense.
▶ It should be consistent with the available information and data.
▶ It should be simple.
▶ It should be capable of being tested, using techniques and materials which are available.
▶ It should be capable of being *either* accepted or rejected when tested. There should be no 'intermediate' decisions.
▶ It need not be 'correct'. If it is not, it can be rejected.

EXAMPLE OF HYPOTHESIS FORMULATION

Using our 'nettle' example, here is a fairly simple hypothesis formulation ('construction') in a biological investigation:

| 1 | Observation 1 |

Plants which receive more light seem to grow bigger.

| 2 | Hypothesis 1 |

The mean (average) height of plants growing in shady conditions will be less than the height of the same type of plant growing in unshaded conditions.

| 3 | Observation 2 |

The mean heights of the shaded and unshaded plants were 69 cm and 55 cm respectively. *Hypothesis 1 is rejected.*

| 4 | Hypothesis 2 |

Shaded plants 'compensate' for reduced light levels by growing taller, so increasing the mean number of leaves per plant.

| 5 | Observation 3 |

The mean number of leaves per plant on shaded and unshaded plants was 16.0 and 13.6, respectively. *Hypothesis is accepted.*

Depending on the time and facilities available, this 'stepwise' process could continue in an almost infinite way. However, you should remember that there will be a corresponding increase in results to record and interpret!

In the example above, the two main types of observation are shown:

- **Qualitative observations** – these involve *descriptions*, e.g. 'plants which receive more light seem to grow bigger'.
- **Quantitative observations** – these involve *measurements*, e.g. 'the mean height of the shaded and unshaded plants were 69 cm and 55 cm respectively'.

Note that both types of observation are useful in experimental work.

The example described here involves the variation of a factor (plant growth) in two distinct situations (sun and shade). However, more able students may be asked to provide several (i.e. at least three) hypotheses to explain a fairly complex situation. These hypotheses are likely to be related. For instance, in a comparison of invertebrate populations in different ponds, differences were found in the population of the freshwater shrimp *Gammarus*. Students suggested several possible hypotheses for this:

- that relatively more *Gammarus* are found in water which has a higher pH
- that relatively more *Gammarus* are found in water which has a lower temperature
- that relatively less *Gammarus* are found in water which is slower-moving.

4.2 Testing a hypothesis

Once a suitable hypothesis has been formulated, the next step is to test it. This will involve many of the practical techniques already outlined in Chapter 2, Units 3 and 4. Having conducted the investigation, it may be immediately apparent that the hypothesis can be rejected or accepted. However, results are not always clear-cut, and information will need to be recorded (Ch. 2, Unit 5) and interpreted (Ch. 2, Unit 6) before this decision can be taken.

Students sometimes feel that their hypothesis should be 'correct' and are sometimes disappointed or frustrated when an experiment does not 'work'. In formulating a hypothesis, you should be ready to reject it, depending on the results. This should not be regarded as a failure!

In the 'nettle' example above, the series of Observation-Hypothesis-Observation- etc. seem to be part of a 'trial and error' process. In a sense, this is true of all science. The investigator should ideally have an 'open mind' about the possible outcome of the investigation. It *is* difficult to plan and conduct an investigation without having a good idea about what might happen. However, some flexibility of thinking is important in interpreting and evaluating an investigation (see Unit 5 below). It is also necessary in formulating further hypotheses. Investigations often begin with ignorance and end with knowledge. One famous biologist once said, "the advantage of a certain amount of ignorance is that it keeps you from knowing why what you have just observed could not have happened"!

UNIT 5 EVALUATING THE EXPERIMENT AND SUGGESTING IMPROVEMENTS

Evaluating an experiment and suggesting improvements may actually be the *starting* point for many investigations. For instance, the evaluation of an experiment originally organised by the teacher (see Ch. 2) may lead to suggestions for further work. This could be the basis of a student's own investigation. Evaluation is an important part of the overall process of experimental design because it can allow the original hypothesis to be accepted or rejected. If necessary, a new hypothesis can be proposed, too. Evaluation is therefore part of a continuing process (Fig 3.6).

The evaluation of an investigation really consists of deciding whether the hypothesis was valid or not. If this decision is difficult or impossible to make, then the hypothesis was probably not formulated correctly (see Unit 4.1). In this sort of situation, it may be necessary to reject the original hypothesis and formulate another.

Another possibility is that, whilst the original hypothesis is accepted, the information gained from the investigation is rather limited or unclear. For example, in the 'nettle' investigation, one group of students suggested that 'leaf number' was not a very useful indication of a plant's photosynthetic ability. The group therefore formulated another hypothesis: 'The mean leaf area of shaded nettle plants will be more than the mean leaf area of unshaded nettle plants'. On measuring leaf areas (see Fig 2.19), they found that the mean areas of shaded and unshaded leaves were 54.3 cm^2 and 44.3 cm^2 respectively. The hypothesis was therefore accepted.

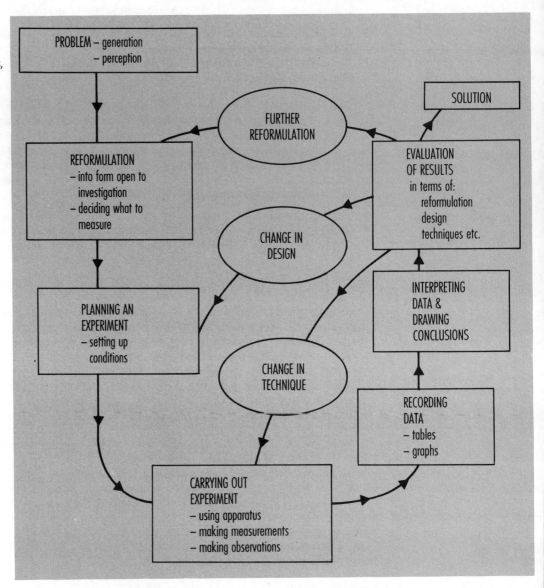

Fig 3.6
Flow diagram, showing the role of evaluation in experimental design, (Reprinted from the IGCSE Syllabus, with permission).

Students may find that, though the original hypothesis seemed quite reasonable, the techniques or materials used in the investigation were unsuitable. Another possibility is that the results were too complex or numerous to analyse properly. These problems show difficulties with the experimental design. However, the experimental design should be based on the hypothesis. This suggests that the experimental design was not appropriate for the original hypothesis. In suggesting improvements, you may need to propose a new hypothesis or experimental design, or both.

This is an important approach in science, and you should not feel a sense of failure because of it. Full credit is given for a good evaluation, even if it reveals that the investigation was a 'disaster'!

Suggestions for improvements means:

> 66 It is important not to spend too much (or too little!) time on your investigations 99

- identifying and eliminating misleading or irrelevant aspects of the original investigation, and
- suggesting more useful and appropriate aspects to be included in any follow-up investigations.

This process is similar to that of experimental design (Unit 3.1), but should be easier because the 'structure' of the design is already available. Hopefully it is the *details* of the design which need attention and not the design as a whole. One difficulty (or delight) in science is that the process of testing-evaluating-improving investigations is limitless! For this reason students who are (a) anxious to be given sufficient credit for their work, or (b) increasingly interested in their investigation, may need to terminate this process 'prematurely'!

5.1 How are marks awarded?

Fig 3.7 shows how the skill of experimental design and problem solving is assessed.

Fig 3.7
'Experimental design and problem solving': skill differentiation.

Standard:	LEAG	MEG	NEA	SEG	WJEC	IGCSE	NISEC
High	Formulate a hypothesis. Choose appropriate methods, with techniques clearly understood and explained. Consider validity, including repetition, controls. Workable, well set out plan. Awareness of limitations in procedure. Make constructive suggestions for improving method or hypothesis. Provide useful suggestions for further work.	Identify problem and outline plan accurately, clearly and without assistance. Select techniques, apparatus and materials confidently and without assistance. Organise and carry out investigation systematically. Evaluate methods and make positive suggestions for improvement spontaneously.	**Skills available:** *Skill 27* Identification of an uncontrolled variable. *Skill 28* Suggestion of an appropriate method of control *Skill 29* Formulating a hypothesis when the data involves a single factor varying in two distinct situations *Skill 30* Formulating several hypotheses in a novel and relatively complex situation involving factors which may interact *Skill 31* Specifying apparatus for the experiment. *Skill 32* Planning the sequence of the experiment	Formulate a hypothesis without assistance. Design an experiment without assistance.	Formulate a relevant hypothesis and design and conduct an experiment which tests it, *without any assistance*.	Identify problem and outline plan accurately clearly and without assistance. Select techniques, apparatus and materials confidently and without assistance. Organise and carry out investigation systematically. Evaluate methods and make positive suggestions for improvements spontaneously.	Decide, and clearly state a relevant problem. Propose a relevant hypothesis. Decide what to measure or observe. Select, modify or build appropriate apparatus. Show need for controls. Evaluate the result and method and, where appropriate, change the technique or design.
Mid	Translate a stated hypothesis into aims with limited assistance. Select method and suitable techniques with some assistance. Show some awareness of validity. Explain some steps in plan. Indicate understanding of limitations in procedure. Have some ideas for improvements. Indicate possibilities for extending the study.	Identify the problem with some assistance but plan the investigation with a minimum of assistance. Choose techniques, apparatus and materials with occasional help. Conduct the investigation with some assistance. Evaluate methods and suggest improvements with some assistance.		Formulate a hypothesis with some assistance. Design an experiment with some assistance.	**Either** Given a hypothesis, design and conduct an experiment which tests it *without assistance* **or** Formulate a relevant hypothesis, and design and conduct an experiment *with some assistance*.	Identify the problem with some assistance but plan the investigation with minimum assistance. Choose techniques, apparatus and materials with occasional help. Conduct the investigation with some assistance. Evaluate methods and suggest improvements with some assistance.	Following consultation, decide what to measure or observe. Select apparatus from alternatives given. Control one/two variables consistently. Evaluate the method and result, and draw appropriate conclusions.
Low	Formulate overall plan of investigation only with assistance. Propose suitable methods only with guidance. Limited attempt to: use a reasonable number of samples, use repeated observations, or include controls. Limitations of procedure appreciated with guidance. Suggest alternative methods or improvement with assistance; limited awareness of further work possible.	Appreciate the nature of the problem and plan the investigation given considerable asisstance. Decide upon choice of techniques, apparatus and materials, organise and conduct investigation given considerable assistance. Evaluate methods and make positive suggestions for improvements given considerable assistance		Formulate a hypothesis with considerable assistance. Design an experiment with considerable assistance.	**Either** Given a relevant hypothesis, design and conduct an experiment which tests it only *with considerable assistance* **or** Formulate a relevant hypothesis, but not design and conduct an experiment even *with considerable assistance*.	Appreciate the nature of the problem and plan the investigation given considerable assistance. Decide upon choice of techniques, apparatus and materials, organise and conduct investigation given considerable assistance. Evaluate methods and make positive suggestions for improvements given considerable assistance.	Given cues, decide what to measure or observe. Select apparatus from alternatives given. Control one variable (not necessary consistently). Make appropriate conclusions.

UNIT 6 IDEAS FOR INVESTIGATIONS

The main sources of information are:

Teacher suggestions

Teachers are likely to make very useful suggestions because they are familiar with individual students' ability, the availability of apparatus and other resources, the time available, etc.

'Spin-offs'

These originate from teacher-organised practical work, including laboratory experiments, fieldwork exercises. Spin-offs may 'suggest themselves' to students during the Biology course. It is an idea for students to keep a rough list of such ideas *during* the earlier part of the course. Experimental design assessments are often carried out towards the end of the assessment period.

GCSE Biology textbooks

Biology books in general p. 7 and practical books in particular p. 43 are likely to be rich sources of ideas for possible investigations. This book provides lists of practical work suggested by the Examining Groups at the end of each skill section (Ch. 2 Units 2–6); these are also included of course in the syllabuses themselves. Each 'student coursework' Unit in Chapter 4 also concludes with suggestions for further work. However, not all of these experiments will be suitable for 'experimental design'.

Public 'information services'

> **Don't forget your parents too! They may be able to suggest some excellent ideas for investigations!**

You may be able to obtain background or source material from various public organisations and also from some companies. Addresses for some of these organisations are given in the 'Sources of information' sections in the Longman 'Revise Guide: GCSE Biology'. Examples of public organisations include *public libraries* and *museums;* their staff are usually keen to help GCSE students in obtaining information for project work etc.

For example, one Examining Group suggests an investigation into the causes of heart disease. Differences in the incidence of heart disease between two groups may be related to any of the following: occupation, diet, hardness of water and exposure to toxins. Data relating to factors of this sort may be provided by the teacher, and students asked to formulate a suitable hypothesis (an actual investigation would probably be very difficult, however).

More able students may wish to seek such information from the sort of organisations referred to above. The suggestions at the end of this Unit (see below) are specifically intended as examples of possible student investigations. You can find further ideas at the end of each chapter in the Longman 'Revise Guide: GCSE Biology'.

6.1 Examples of practical work involving this skill

The following list of examples, suggested by Examining Groups, are given as *observations*. Each of these ideas could be the basis of an investigation. Before the investigation can be planned, however, it is necessary to (a) identify the problem (Unit 2), (b) formulate a hypothesis (Unit 4), and (c) design a suitable experiment (Unit 3).

- A person's pulse rate changes from 70 to 120 beats per minute after exercise.
- Fruit juice stays fresher for longer if it is kept at lower temperatures.
- The leaf size of nettle plants growing in shady conditions is different from those growing in sunny conditions.
- Pieces of potato shrink by different amounts when they are placed in salt solutions of different strength.
- The concentration of reducing sugar in a solution determines the colour produced when the Benedict's test is performed.
- Heather plants in exposed situations grow slower than those growing in sheltered situations.
- Plaque is more common when the diet has a relatively high sugar content.
- Milk stored in a fridge stays fresh for longer compared with milk stored at room temperature.
- Auxins sold as rooting compounds promote the development of the roots of geranium plants.
- The relative numbers and types of organisms living in two different rock pools are different.
- Cake is more fattening than bread.
- The rate of ventilation in locusts is increased when they are kept in a more confined space.
- There is a difference in the growth of plants on either side of a hedge.
- Cucumber leaves do not grow at a uniform rate.
- Dough rises more quickly if it is kept in a warm kitchen.

STUDENT COURSEWORK WITH EXAMINER COMMENTS

UNIT 1 INTRODUCTION

This chapter is composed of typical student written coursework, covering a very wide range of subject material. The coursework has been selected because it is representative in several ways:

- It covers most aspects of the GCSE syllabuses.
- It includes all of the 'written' skills by direct example, with references to 'practical' skills, too.
- It includes a variety of different assessment standards.

Each piece of coursework includes a description of the experiment or exercise, a list of the skills being assessed, examiner comments, and suggestions for further related work. Each Unit is sub-divided into the following sections:

Background

This briefly explains the theoretical and practical context of the sample coursework. The background often includes general comments of the frequency of particular types of coursework. Some description is also given of the instructions originally given to the student.

Cross-references

The samples of coursework are arranged in units which correspond with the chapters of the Longman 'Revise Guide: GCSE Biology'. This allows the reader with access to that book to make *cross-references* to the more detailed theoretical background of each piece of coursework. Cross-references to the 'Revise Guide' book are abbreviated throughout this chapter as *RG X-ref*. The intention is to reinforce the complementary nature of practical and theoretical aspects of Biology.

Typical student's coursework

Samples of coursework are based on genuine work, but some changes have been made for the sake of clarity and also confidentiality. The standard of the work is quite variable but, generally it is of a moderate to high standard.

Examiner comments

The examiner comments are made to show the reader aspects of the coursework which have been particularly well done, and aspects which could have been improved upon. Examiners and moderators do not normally write comments on the coursework itself.

Possible skills assessed in coursework

The actual skills assessed will depend on the nature of the coursework, the Examining Group under which the assessment is taking place, and the student's current ability in the skills involved. The titles of the various skills are those used in Chapters 2 and 3, and cross-references are given. Note, however, that the different Examining Groups refer to the skills (or skill areas) in different ways (see Fig 1.4).

Individual pieces of coursework may involve a wide range of skills. However, any particular assessment will normally be limited to perhaps one or two of the most appropriate skills. These may include 'practical' skills (e.g. 'Handling apparatus and materials') which obviously are not (directly) apparent in the written accounts presented in this chapter.

Examiner assessment

A broad assessment of the overall standard is given for each example of coursework. Only the 'written' skills which are evident in the sample are assessed here. Actual marks are not given for these samples because there is considerable variation in practice between Examining Groups. In any case the original assessment may have been based partially on 'practical' skills which are not included in the written samples shown here.

Suggestions for further work

Suggestions are made for further work that might be conducted on the piece of coursework and also for *related* and *linked* work. These will all probably involve using additional techniques, for instance in 'practical' skills or in presentation. These suggestions are provided:

▶ to prepare you for a range of *teacher-organised* practical work that might be expected, and
▶ to provide ideas for possible *student-organised* investigations (see Chapter 3).

UNIT 2   ESSENTIALS OF LIFE – ENZYMES

Background

Enzyme experiments are very common in GCSE Biology, and make a popular choice by teachers for coursework assessments. There are several reasons for this. For instance, enzyme experiments provide an excellent opportunity for students to demonstrate their manipulative abilities. A very common format is the effect of *environmental condition,* often temperature or pH, on the *rate of enzyme activity.* Enzyme activity can be monitored either by the rate at which substrate is used up, or by the rate at which the product is formed.

This experiment is intended to demonstrate the presence of a starch-digesting enzyme in saliva. The enzyme in fact is amylase (diastase). This experiment could be readily developed into a more complex investigation. *RG X-ref.* pp. 17–18.

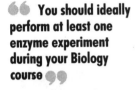

❝ **You should ideally perform at least one enzyme experiment during your Biology course** ❞

Fig 4.1
A typical enzyme experiment

CHAPTER 4 ESSENTIALS OF LIFE – ENZYMES

Title: To show the effects of saliva on starch

> 66 This is a good way of showing the experimental set-up. Larger, neater diagrams preferred 99

> 66 Room temperature could be *measured* and recorded 99

> 66 How much is 'a drop'? Was the amount the same each time? 99

> 66 What is the iodine used for? 99

> 66 Column heading omitted 99

> 66 It might be an idea to write down the contents of each tube here; to be referred to later 99

> 66 'Black' here suggests incomplete mixing of sample + iodine 99

> 66 Where is the result for 16 mins? *Care* is needed when recording/copying results! (This is a critical stage in the experiment) 99

> 66 It is a good idea to refer directly to actual results; e.g. time/colours 99

> 66 What does this mean? This seems to be an incomplete reference to 'theory' 99

> 66 This might be an opportunity to refer to the products of digestion (i.e. reducing sugars) 99

> 66 Confusion here. It is the *iodine* which changes colour 99

> 66 Molecules are not 'killed'! The enzyme is *denatured* ('destroyed') by heating 99

> 66 This is a good suggestion. 99

Method

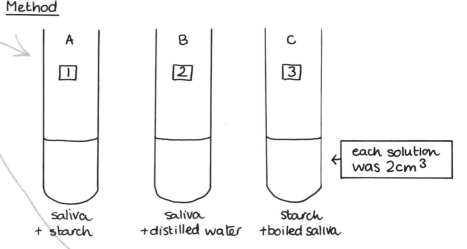

saliva + starch | saliva + distilled water | starch + boiled saliva

each solution was 2cm³

We set up the three tubes as shown in the diagram. The solutions were all mixed at the same time, and then left at room temperature.

We took a drop of each mixture every 2 minutes, starting immediately. The drops were added to drops of iodine which were already in a 'dimple tile'. We continued for about 20 minutes.

Results

	Tube 1	Tube 2	Tube 3
0	black	orange	black
2	black	orange	black
4	black	orange	black
6	brown	orange	black
8	brown	orange	black
10	black	orange	black
12	brown	orange	black
14	orange/brown	orange	black
18	orange	orange	black
20	orange	orange	black

Conclusion

There was a change in colour in Tube 1 but not in the other two tubes. This is because Tube 1 contained both saliva and starch – they are both needed for 'digestion'. Saliva contains the enzyme amylase, which breaks down starch. – This is shown by the colour change.

Tube 3 contained boiled saliva and didn't cause the starch to change colour. This is because the enzyme in saliva is killed by the boiling and so doesn't work. Tube 2 is for comparison. The enzyme took 18 minutes to change all the starch.

Criticisms

This exp. seemed to work well, as there was a colour change in the tube that was supposed to change. For improvements, I would suggest using different temperatures to see the effect of this on the enzyme.

Possible skills assessed in this coursework

'Handling apparatus and materials' (Ch. 2, Unit 2), 'Observing and measuring' (Ch. 2, Unit 4) 'Recording and Communication' (Ch. 2, Unit 5), 'Interpretating information' (Ch. 2, Unit 6).

Examiner assessment

Moderate standard piece of work. The student's account is rather brief and some good observations are not very fully developed. The interpretation could have been more carefully organised, and more direct comparisons made to the different tubes, including the controls.

Suggestions for further work

This experiment: Effects of temperature, pH. Testing for reducing sugars (Benedict's solution). Comparison of saliva from smokers/non-smokers.
Related practical work: Other suitable enzymes (and substrates) include pepsin, lipase, catalase.
Linked practical work: Food analysis (Unit 4.17).

UNIT 3 ESSENTIALS OF LIFE – CELLS

Background

66 **Refer if necessary to pp 13–14 on the use of the microscope** 99

The use of the microscope and the examination of cells is a required assessment for some Examining Groups. The 'cell concept' is a fundamental aspect of Biology because all organisms are made up of at least one cell. Many important organisms (e.g. bacteria, Protista) are too small to be seen without a microscope. The use of the microscope is an important manipulative skill (Ch. 2, Unit 3.3), and allows *identification* and *comparison* of different cells. The contents of smaller cells can, however, be difficult to see; it is important for pupils to record only what they *can* see!

Fig 4.2
Separating onion epidermis

Onion epidermis is only one cell-layer thick, yet can be easily separated (Fig 4.2). If a *small* piece (i.e. about 4 × 4 mm) is carefully mounted on a microscope slide, the individual cells can be clearly seen. This example of coursework is extended into a comparison between cells mounted in distilled water and cells mounted in 20 per cent sucrose solution. *RG X-ref.* p19, pp 21–24.

Title: Comparison of onion epidermis cells in two different solutions

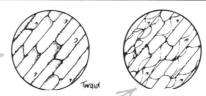

Turgid

FEATURE	CELLS IN DISTILLED WATER	CELLS IN 20% SUCROSE SOLUTION
shape	Fairly regular, box-like	cell walls curved inwards
arrangement	tightly-fitting	loosely-fitting
nucleus	present in all cells	present in all cells
cytoplasm	around the edge of each cell	pulled away from each cell wall, towards the centre of the cell.
state	turgid	plasmolysed
cell membrane	intact	disrupted
vacuole	fluid filled	shrunken

66 Excellent standard of drawing. Cytoplasm should be shown *clear*, unless it can be seen to have the 'granular' appearance suggested here 99

66 What is the scale of these drawings? *Magnification* should be stated 99

66 This is *similarity*; other features are *differences*. It is a good idea to separate these into two groups 99

66 These are descriptive terms, rather than visual features 99

66 This feature is not actually *visible* in the drawings 99

CONCLUSION
Cells take part in osmosis. This is because they have selectively-permeable membranes. Osmosis is the movement of water molecules from a place where their concentration is high to a place where their concentration is low, through a selectively-permeable membrane. This movement is along a concentration gradient. In this experiment the concentration gradient goes in different directions in each situation.

DISTILLED WATER
The highest concentration of water is <u>outside</u> the cells, water therefore moves into the cells, causing them to swell up and become turgid.

SOLUTION
The highest concentration of water is <u>inside</u> the cells, so water moves out from them. This results in them 'collapsing' and becoming plasmolysed.

66 Credit is not given for 'theory' in coursework, but it may be very helpful in the actual *interpretation* of results 99

66 The student has demonstrated a very good understanding here; the different interpretations for the two situations are well-presented. 99

CRITICISMS

1. I could have examined more cells, to be sure that I didn't have 'unusual' results.

2. I could have stained the cells with iodine.

3. I could have used a range of different cells to show 'incipient plasmolysis'

66 A good point. Considerable variation often occurs between cells immersed in the same solution 99

66 Why? This needs explaining! 99

66 This is a good suggestion, but the concept of 'incipient plasmolysis' is probably beyond GCSE and in any case needs explaining! The sketch-graph is a good idea. 99

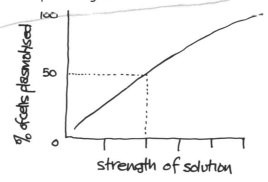

Possible skills assessed in this coursework

'Handling apparatus and materials' (Ch. 2 Unit 3), 'Observing and measuring' (Ch. 2 Unit 4) 'Recording and communication' (Ch. 2 Unit 5), 'Interpreting information' (Ch. 2 Unit 6).

Examiner assessment

Generally *high standard* of work. The high quality of drawing is supported by a good interpretation, though it is rather 'theoretical', with limited reference to actual observations.

Suggestions for further work

This experiment: use of a range of different sucrose solutions; test to see if plasmolysis is reversible, by irrigating the slide (see p 13) with distilled water.
Related exercises: examine cells from other 'higher' plants (e.g. from rhubarb petiole epidermis) and 'lower' plants (e.g. algae such as *Pleurococcus, Cladophora*; also, moss leaf cells).
Cheek cells are very convenient examples of animal cells (though some schools/LEAs may have restricted their use because of concern with the spread of HIV). Various Protista can be found in still ponds, murky farmyard puddles etc.
Linked work: Osmosis (below), Classification (Unit 5).

UNIT 4 ESSENTIALS OF LIFE – OSMOSIS

Background

Osmosis is an important characteristic in cells because it depends on the presence of a selectively-permeable membrane. Osmosis can be demonstrated in a wide variety of practical exercises. Many of these are relatively simple to perform, though they require careful observations and interpretation.

'Osmosis' practicals are *very* common in coursework assessments and often involve the effect of at least two concentrations of salt or sucrose solution on tissues. The effect can be observed at the cellular level (Unit 3, above) or at the tissue level. Changes to tissues can be monitored by measuring changes in dimensions (length, volume), mass or shape. The 'turgidity' of the tissue can also be determined by feel.

This example of coursework involves the study of osmosis in potato cylinders; these are cut to a uniform size and shape, making length measurements easier. The cylinders are immersed in a range of different concentrations (Fig 4.3) for the same period of time, then re-measured. *RG X-ref.* pp 21–24, p 28.

> 66 **Many teachers use this exercise (or similar ones) for assessment practicals** 99

Fig 4.3
Potato cylinders immersed in different concentrations of sucrose

CHAPTER 4 ESSENTIALS OF LIFE – OSMOSIS

Title: To study the effect of different concentrations of sucrose on potato rods

Fig 4.3a

Method.

Each group cut 4 rods of potato using a cork borer. The rods were trimmed by a scalpel so that they were all 30mm long. Then we put them into 4 different sucrose solutions.

We left the rods for 30 mins. then measured them again. We also felt the potatoes before and after. The results for all the class were written on the board and we copied them down. The results are shown in the table.

Results.

Solution	time: 0 mins	time: 30 mins	'feel'	averages
0%	30	33 32 34 / 33 32 33	v. hard (turgid)	32.8 / +10%
5%	30	30 29 30 / 28 30 28	fairly hard	29.2 / -3%
10%	30	26 26 27 / 21 27 26	soft, rubbery	25.5 / -13%
20%	30	25 26 23 / 26 24 23	v. soft (flaccid)	24.5 / -17%

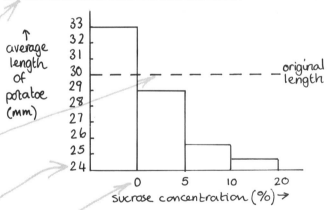

Graph to show class results for potato exp.

Conclusion

This experiment showed that potatoes change in length according to the solution they're in. The length got shorter as the strength of the solution got more. You could also feel a change in them. As the solution increased, the potato got softer. We calculated the percentage change in length and used that for the graph above. This shows that they increase in distilled water and decrease in the sucrose solutions. This is due to osmosis.

Left margin comments:

- This length may have been given in the teacher's instructions. However, students should note that measurements are likely to be more accurate using lengths of 40 mm or even 50 mm.
- The actual concentration should be given. (Also, how were they prepared)
- A more complete heading needed, e.g. 'Concentration of sucrose solution (%)'
- These two columns could be presented more clearly! For instance, 'Length (mm) of potato rods' written over both columns. Each column could then have separate headings, e.g. 'before immersion', 'after immersion (class results)'.
- Titles for graphs are important; however, this one could be more informative!
- This line is a good idea; it allows comparisons between lengths 'before' and 'after' immersion
- This is an excellent example of the use of scales, from 24 mm rather than from 0 mm!
- These values refer to the 'blocks' on the graph, and would be better written centrally under each block

Right margin comments:

- This column seems to have been added on as an afterthought! This could be improved by drawing separate columns for 'averages' and 'percentages'. The calculations for the percentages could be shown, and some reference to the values made in the text!
- This value seems low in comparison with others at this concentration. Students have the option to omit such values in the average (i.e. mean) calculations. In this example, a mean value calculated on the other five values would be 26.4.
- It is a good idea to state the relationship between sucrose concentration and potato length in this sort of way. In this example, the statement does not apply to the distilled water (0%)
- Direct references to the actual percentage values would have been very useful. For instance, 'there was a 13% difference in length between rods at 0% compared with 5% sucrose'
- The experiment allows the pupil an opportunity to provide a more complete explanation of the observations in terms of osmosis. The student shows an awareness of the importance of osmosis, but seems reluctant to be more committed than that!

Possible skills assessed in this coursework

'Observing and measuring' (Ch. 2, Unit 4), 'Recording and communication', (Ch. 2, Unit 5), 'Interpreting information' (Ch. 2, Unit 6).

Examiner assessment

Moderate standard overall, though the quality of the written account is quite variable. The observations seem fairly consistent and results are quite well presented. If students have received help with these aspects of the practical work the teacher would be aware of this during the assessment.

The histogram drawn by the student is only one way in which the results might have been presented. A good *line graph* would show whether changes of length occurred in a regular way with changing sucrose solution. It would also allow the concentration of cell contents to be estimated (where there is no gain or loss in length) by interpolation (see Ch. 2, Unit 6).

In this example, the student provides a fairly limited interpolation of the results. The following points could have been referred to in the interpretation:
i) comments on the variation within the class results – possible causes for this include inconsistent techniques of measurement (of length or time) and incomplete immersion of potato cylinders
ii) an explanation of the results in terms of osmosis (e.g. 'exosmosis' and 'endosmosis'); cross-reference between the measurements (length) and the observations ('feel')
iii) criticisms of the experiment and suggestions for further work (see below).

Suggestions for further work

This experiment: Changes in volume and/or mass also measured.
Related experiments: Observations of cells (Unit 3), observations of the colour leached out of washed beetroot tissue.
Linked work: Cells (Unit 3).

UNIT 5 — DIVERSITY OF LIFE – NATURAL CLASSIFICATION

Background

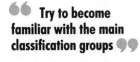

66 **Try to become familiar with the main classification groups** 99

Natural classification involves placing organisms into groups. These groups are occupied by different species ('types') of organisms which share similarities, for instance in body form and structure. Each species is given scientific names according to which classification groups (including 'species') they belong to. Organisms belonging to a particular species have many features in common. This fact can be used to *identify* an unknown organism (see Unit 6) on the basis of visible characteristics.

In GCSE Biology, the emphasis in classification is on the relationships between species as revealed by their similarities. Students will not be expected to provide a detailed identification of specimens. However, some awareness of the major classification groups and their significance will be useful. For instance, Figure 4.4 shows a locust and a cockroach. They are both insects, although they belong to different species. The insects are a biologically important group, and these two species are also important because of their impact on human populations.

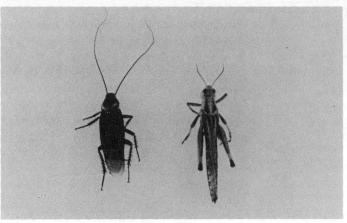

Fig 4.4
Example of two insect specimens

CHAPTER 4 DIVERSITY OF LIFE – NATURAL CLASSIFICATION

In this sample of coursework, the student was asked to (i) draw the specimen, (ii) comment on structures which may adapt the organism for its type of life, and (iii) comment on the possible classification of the specimen. *RG X-ref.* pp 31–37

Title: Characteristic and adaptive features of 'Specimen A'

Fig 4.4a

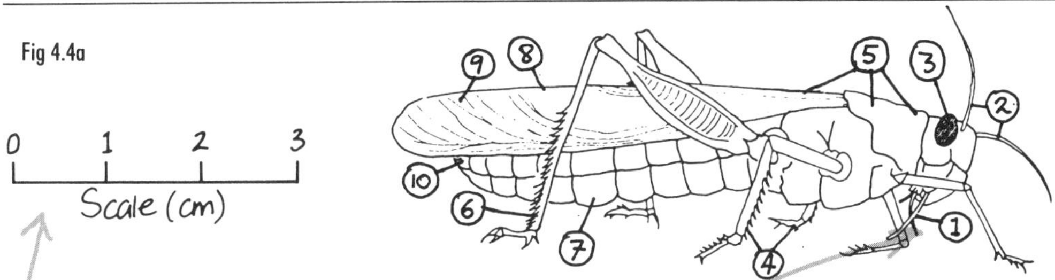

Scale (cm) 0 1 2 3

> **An excellent standard of drawing, with correct proportions. Some details (e.g. the mouthparts) are inaccurate, however.**

> **Good. It is important to include an accurate scale with all biological drawings**

> **Separate, more detailed drawings can be included of structures which are small on the main diagram.**

> **The numbered cross-references to the drawing are a very good idea. An alternative method of presenting the information in this table would be as notes ('annotations') arranged around the drawing itself**

> **Good. These adaptations are mainly based on actual observations, visible on the drawing. However, some of the observations could have been developed further**

> **This is not a structural feature, though the student may have observed this in a living specimen in the laboratory**

> **(Credit is given for the *observations* here and not for knowing the names of structures)**

Adaptations of 'Specimen A'

Structure/feature	Adaptation
① mouth parts	these perform different functions, like grasping, chopping, piercing etc. (the food is mainly grass)
② antennae	these provide the animal with 'touch' information about its surroundings, including food and other animals.
③ compound eye	this helps the animal to locate food etc.
④ jointed legs	these allow the insect to move along stalks of grass, cereals etc. The back legs allow a 'jumping' action, to get rapidly from place to place
⑤ body parts	there are three main parts to the body: they are called head, thorax and abdomen
⑥ spines	the spines on the back legs are for making a noise, by rubbing legs together. This allows communication between individuals
⑦ spiracles	these allow the animal to 'breathe' by drawing air in and expelling it again.
⑧ wings	allows the insect to fly from one area of food to another
⑨ colour	the 'mottled' browny colour is camoflage, this enables the insect to 'blend in' with its background and not be seen by predators
⑩ reproductive parts	these allow the insect to mate and have young

Classification of 'Specimen A'

<u>Animal</u>. It has the characteristic form of an animal, with legs, sensory organs etc. It feeds by consuming food.

<u>Invertebrate</u>. It has an <u>exoskeleton</u>.

<u>Arthropod</u>. It has jointed limbs.

<u>Insect</u>. It has three pairs of <u>jointed limbs</u>, <u>three regions to the body</u> (head/thorax/abdomen), <u>two pairs of wings</u>.

> **This is not characteristic of invertebrates in general. Also, it is not referred to in the drawing or table!**

> **A good, distinguishing feature**

> **Good**

Possible skills assessed in this coursework

'Observing and Measuring' (Ch. 2, Unit 4), 'Recording and communication, (Ch. 2, Unit 5)

Examiner assessment

High standard of drawing and of most observations. The student has obviously examined the specimen and thought with care about adaptations.

Suggestions for further work

This experiment: Observations of behaviour in living specimens.
Related exercises: Comparisons with other specimens, e.g. insects, other arthropods.
Linked work: Artificial classification (Unit 6), Development (Unit 10).

UNIT 6 DIVERSITY OF LIFE – ARTIFICIAL CLASSIFICATION

Artificial classification is used as a means of distinguishing different species ('types') of organisms from each other. This is done by identifying *unique characteristics,* usually *visible.* These allow 'unknown' specimens to be placed into groups, perhaps even the species group. Note, however, that the distinguishing features are used simply for sorting specimens into groups. The features may not necessarily be biologically significant. The sorting process often involves asking a series of paired alternative questions about the specimen. A series of such questions is called an *identification key.*

In this coursework sample, the student was asked to collect six different tree leaves (Fig 4.5). The task was then (i) to identify distinguishing features, and (ii) construct an identification key. *R.G. X-ref:* pp 36–37.

> **The use of keys is expected by some Examining Groups**

Fig 4.5
Six different tree leaves: A = elm, B = field maple, C = horse chestnut (part of compound leaf), D = beech, E = lime, F = sycamore

CHAPTER 4 DIVERSITY OF LIFE – ARTIFICIAL CLASSIFICATION

Title: Making an identification key

Fig 4.5a

Key of tree leaves

(i) Distinguishing features

LEAF	long	small	rounded	toothed edges	long stalk	variegated
A	x	✓	x	✓	x	x
B	x	✓	✓	x	✓	x
C	✓	x	x	✓	x	x
D	x	✓	✓	x	x	x
E	✓	x	✓	✓	✓	x
F	x	x	x	✓	✓	✓

Fig 4.5b
Identification key
for the six leaves
shown in Fig 4.5

(ii) Identification key

1. Is the leaf small?   Yes .. go to 2. (A,B,D).
                              No .. go to 3. (C,E,F).

2. Is the leaf rounded?   Yes .. go to 3. (B,D).
                              No .. it is A

3. Is the leaf toothed?   Yes .. go to 5. (C,E,F).
                              No .. go to 4. (B,D).

4. Has the leaf got a long stalk?   Yes .. it is B
                              No .. it is D

5. Is the leaf variegated?   Yes .. it is F
                              No .. go to 6. (C,E).

6. Is the leaf long?   Yes .. it is C
                              No .. it is E

> **This is a good, systematic way of presenting distinguishing features. The number of features needed will normally be *at least* one less than the total number of specimens; in this case, 6–1 = 5. Each horizontal row should have a unique array of ticks and crosses**

> **'Small' is a difficult feature to apply, for instance, because of age differences. Also, 'small' depends on individual judgement.**

> **'Rounded', again, is difficult to apply; it's a rather ambiguous feature.**

> **A good feature ('serrated edge' would be another way of expressing this feature)**

> **Not a reliable feature. The length of the stalk may depend on where the leaf was broken off the tree!**

> **A good choice of feature, easy to apply**

> **This is well-organised; the six leaves have been divided into two equal-sized groups. Each group can then be dealt with separately.**

> **Problems here. This question does not actually achieve anything, as neither group is split!**

> **Like 'small', this is rather vague. 'Longer than broad' might be better, as this tends to remain constant during the leaf's development**

> **If the *name* of the specimen is known or is given, it can be written next to the appropriate letter.**

> **An improved (but not perfect!) version of the student's key is given alongside:**

① serrated?   Yes … ② (A,C,E)
                      No … ④ (B,D,F).

② longer than broad?   Yes … = [C]
                      No … ③ (A,E)

③ roughly symmetrical?   Yes … = [E]
                      No … = [A]

④ lobed?   Yes … ⑤ (B,F)
                      No … = [D]

⑤ variegated?   Yes … = [F]
                      No … = [B]

Possible skills assessed in this coursework

'Observing and measuring' (Ch. 2, Unit 4), 'Recording and communicating' (Ch. 2, Unit 5).

Examiner assessment

Moderate standard. The student has demonstrated fairly good observation skills and the key is quite well constructed.

Suggestions for further work

This exercise: Transform the key into a branching 'spider' diagram. Extend the task by (i) including more specimens, (ii) referring to more detailed structure (e.g. veins, surface hairs) and using more 'botanical' descriptions (e.g. 'ovate', 'palmate').
Related work: Practise using ready-made keys to identify a small range of 'unknown' specimens.
Linked work: Natural classification (Unit 5).

UNIT 7 REPRODUCTION – SEXUAL REPRODUCTION: FLOWERING PLANTS

Background

> You may need to revise flower structure and function at this stage

Sexual reproduction occurs in most plants and animals. The process involves structures for the production, transfer and fusion of gametes (sex cells). In higher plants, the organs of sexual reproduction are *flowers*. The main two functions of flowers are

- to allow the transfer of pollen and
- the production and dispersal of seeds.

This piece of coursework is based on a detailed examination of a flower. The student was asked to (i) dissect the flower to reveal its structure (Fig 4.6), and (ii) label and comment on any structures that are characteristic of insect-pollinated flowers. *RG X-ref* pp 49–52.

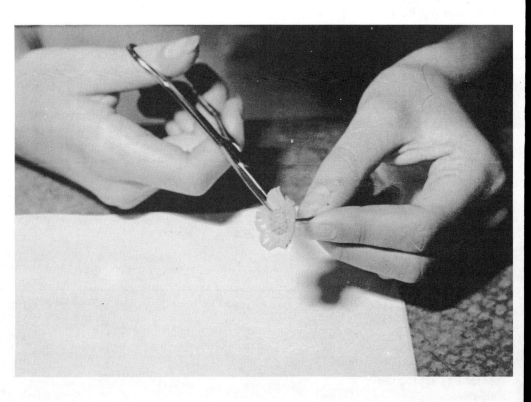

Fig 4.6

Title: Examination of an insect-pollinated flower

> **An excellent drawing.** This shows the main structures in outline and in correct relative proportions. However, no *scale* is given; this may result in less marks being awarded.

> **The details shown** in the drawing do not obscure the clarity of the drawing itself. The details are mostly relevant to the pollination of the flower.

> **Good observation** (it is the filament which is stiff); the pupil should explain the significance of this.

> **Yes; why?** What about the size, shape of the pollen?

> **Good point.** The student should indicate if the drawing is based on more than one flower, at different stages of development

> **No. These are actually veins within the petal**

> **Good;** the importance of this needs explaining

> **Good.** The use of colours in drawings is not generally a good idea. However, it is important to note the colour since it is an adaptation for pollination.

> **A relevant feature.** However, was it visible here? Also, what about scent?

> **This label needs some explanation.** Does it refer to the type of insect likely to visit the flower?

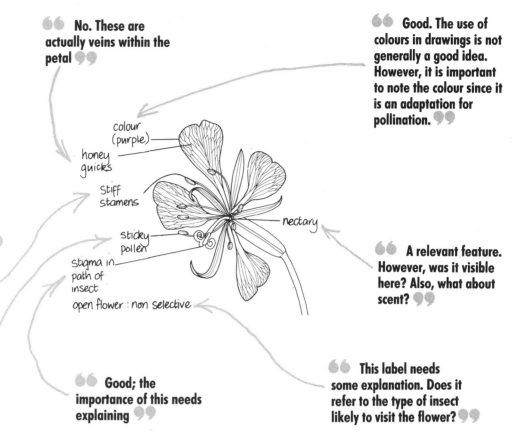

Labels on drawing: colour (purple); honey guides; stiff stamens; sticky pollen; stigma in path of insect; open flower : non selective; nectary

Note: On any single flower the stigma remains closed while the anthers are shedding pollen. It opens out after the stamens start to droop. This prevents self pollination.

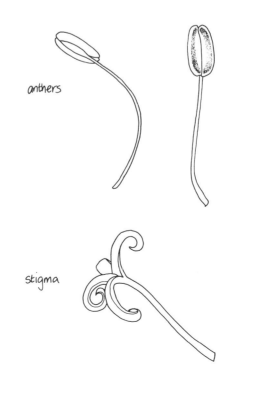

anthers

stigma

> **It is an excellent idea** to provide separate, detailed drawings like this when structures are small.

> **What are the adaptations here, and in the drawing above? Also, what is the *scale*?**

Possible skills assessed in this coursework

'Handling apparatus and materials' (Ch. 2, Unit 3), 'Observing and measuring (Ch. 2, Unit 4), 'Recording and communication' (Ch. 2, Unit 5), 'Interpreting information' (Ch. 2, Unit 6).

Examiner assessment

Moderate to high standard. The standard for different skills shows some inconsistency. This may be because the student is not equally good (or bad) in all the skills involved here. Alternatively, the student may have misunderstood the task. In this sort of situation the pupil might simply retain the marks for the 'best' skill on this occasion.

The quality of the drawings is excellent, showing a very good standard of observation. However, the student has not made sufficient use of these observations! The student could have provided a more detailed *interpretation* of the biological significance of the various structures. In this example of coursework, the student might be given high marks for 'observation and measurement', and fairly low marks for 'interpretation of information'. However, this last skill will be assessed on other occasions, perhaps with better results.

Suggestions for further work

This exercise: More detailed examination using a microscope (low power). Observe the behaviour of pollinating insects.
Related exercises: Compare a range of insect- and wind-pollinated flowers. Germinate pollen grains in 15 per cent sucrose on a microscope slide. Asexual reproduction in plants (i.e. rooting experiments, examination of perennating organs, spores).
Linked work: Natural classification (Unit 5), Seed/fruit dispersal (Unit 8).

UNIT 8 REPRODUCTION – SEED DISPERSAL

Seeds are released at a particular stage in a plant's life-cycle, following pollination and fertilization. Seeds are dispersed in a variety of ways, for instance by wind and by animals. There are two main advantages to successful dispersal:

> *This is an example of a 'model' system in Biology. The model represents a simplification of the real thing*

- *colonisation* by the species of a new area, and
- *avoiding competition* with the parent plant.

Wind-dispersed fruits are adapted in various ways to be carried away from the parent release, for instance by having wing-like extensions. These operate by delaying the descent rate of the fruit.

The coursework illustrated here is an example of a *model system* in Biology. Sycamore trees have winged fruits (Fig 4.7) which fall slowly and erratically to the ground. In the experiment, a model 'sycamore fruit' is constructed and used to study the effect of wing size on descent rate. The advantage of this model is that changes can be made to it very easily. Also, it can be used when wind-dispersed fruits are not available! *RG X-ref* pp 52–53, p64.

Fig 4.7
An example of wind dispersal – sycamore fruits

CHAPTER 4 REPRODUCTION – SEED DISPERSAL

Title: Experiments in descent rate in a model 'sycamore fruit'

Apparatus

paper pattern   metre ruler   scissors
paper clip   accurate stopwatch

> **Credit is not given in coursework assessments for lists of apparatus unless they are part of a student's investigation plan**

Method

We constructed the model sycamore fruit from the paper pattern. When the model was ready, we released the model from a height of 2m from the ground, and timed how long it took to reach the ground. This was repeated twice. Then the length of both the wings was reduced by 10mm and the experiment was repeated. This procedure was carried out for a total of six sets of readings. These are shown in the table in the results.

> **More information could be provided about this**

Results

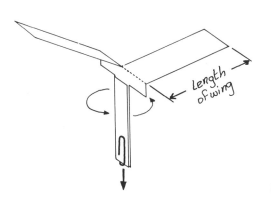

length of wings (mm)	descent time(s) for 2 metres			
	1st attempt	2nd attempt	3rd attempt	mean
100	2.19	2.13	2.06	2.13
90	2.03	1.86	1.80	1.90
80	1.78	1.74	1.82	1.78
70	1.75	1.42	1.38	1.52
60	1.41	1.26	1.41	1.39
50	1.29	1.28	1.19	1.25

> **A very clear and well-organised table of results**

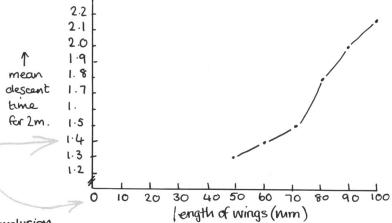

> **A good graph, which illustrates the data well**

> **The accuracy of the graph could have been increased by plotting the x-axis 40 (to 100) mm.**

> **Two what? This could be stated more clearly**

Conclusion

The shorter the wing length, the shorter the descent time. The two are directly related to each other. This means that sycamore fruit which want to be dispersed further will need longer wings.

> **Not very 'biological'! An alternative way of expressing this idea might be: 'Sycamore fruits with longer wings may have an increased descent time and are therefore more likely to be dispersed further'. Note that this is only *suggested* by these results, not *proven*!**

> **Excellent points**

Criticisms

The experiment worked very well. Not all the readings were the same for each wing length, though. This could be because:
a) the descent time varied, e.g. because of draughts in the lab.
b) the timing was not always correct. It was difficult to time accurately when the descent was quick.

Improvements
a) Try this experiment with real sycamore 'helicopters'.
b) Do this experiment with a 'side wind' blowing from a hairdrier.

> **This needs further explanation. How could similar methods be used?**

> **Well, yes. Very difficult to organise, measure and interpret, though!**

Possible skills assessed in this coursework

'Observing and Measuring' (Ch. 2, Unit 4), 'Recording and communication, (Ch. 2, Unit 5)

Examiner assessment

High standard of measurement, though not accompanied by any descriptive observations. The conclusion is *low* in standard, but the criticisms are well-chosen; the overall standard of interpretation is therefore *moderate*.

Suggestions for further work

This experiment: Calculation of descent time in metres per second. Changes to the basic design, including the added mass.
Related work: Comparisons of fruit dispersal in different species. Drawings of various fruits, showing adaptations for dispersal.
Linked work: Growth (Unit 9), Populations (Unit 19).

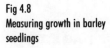

UNIT 9 — GROWTH AND DEVELOPMENT – GROWTH

Fig 4.8
Measuring growth in barley seedlings

Background

Growth is a characteristic of all living things. Growth may be defined as an increase in the living matter within an organism. Growth can be measured by changes in the size or mass of an organism. It can be shown that the rate of growth for many species is not constant within individuals, and also varies between organisms in a population.

This example of coursework is an investigation designed by two students. They made regular measurements of the leaf number and height (above soil level) of a population of twenty barley seedlings (Fig 4.8). The seedlings were germinated at the same time, and subjected to the same conditions throughout their growth period. *RG X-ref.* pp 67–73.

66 Remember that there are several methods for measuring growth 99

Title: Investigation of the growth patterns of barley seedlings

66 A good hypothesis, though 'occurs in a regular way' needs further explanation 99

Hypothesis: That growth of oats plants, measured by (a) leaf number and (b) height, occurs in a regular way.

66 Is this part of a *plan*? An outline of the investigation plan should be included, however 'simple' the investigation 99

Apparatus needed Tray of compost. Ruler. barley seeds

Method The seeds were soaked for 48 hrs before the experiment was due to start, then they were planted in a tray of compost. The tray was left in the prep. room (by the window) throughout the experiment. The seeds were watered 2-3 times each week.

66 More details needed here! Which compost? How deep? Density of planting? 99

Once a week the tray was brought into the lab. and 20 seedlings were measured in the following ways:
(a) number of leaves visible (b) maximum height (in cm).
The experiment was continued for a total of 6 weeks.
A summary of results is shown below.

66 This is not very clear. Were only 20 seeds sown? If so, did they all germinate? If not, how were the seedlings chosen? 99

Results

week number	1	2	3	4	5	6
mean no. leaves	0	1.0	2.3	3.3	4.6	7.1
mean height (cm)	0.2	8.3	16.4	21.4	30.5	41.0

66 Maximum height for each plant, presumably. Is this the height to the top of the longest leaf? 99

66 This is a good summary of the data. However, the *raw data* should also be included, so that the teacher/ examiner can check the calculations. Also, the missing data represents 'lost information', e.g. of *variation* within the population. 99

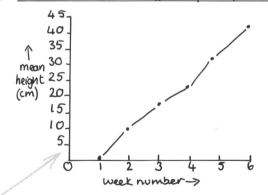

66 An excellent graph, with well-chosen scales. 99

CHAPTER 4 STUDENT COURSEWORK WITH EXAMINER COMMENTS

> **A generally good graph**

> **The y-axis should have been drawn to allow for this maximum value!**

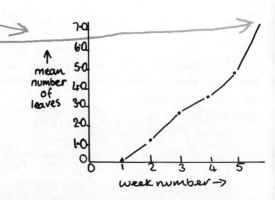

> **Rather a vague comment, though it is followed by something more specific. It is a good idea to refer directly to the data/graphs to support such statements. Each 'part' of the curves on the graphs could, for instance, be labelled 'A', 'B', etc and comments made on each part. Perhaps the growth *rates* could even be calculated.**

<u>Conclusion</u> The graphs show a fairly rapid increase in leaf number and height. The rate for both forms of growth is more or less constant, though it increases more quickly in weeks 4–6. The increase in the number of leaves and also the height are important to the barley plant because it can absorb more light for photosynthesis.

> **Excellent understanding shown**

<u>Comments on the experiment</u> The results from this experiment seemed quite reliable. However, the outcome of the experiment would be more accurate if:
(a) more plants were used, e.g. 20 or more.
(b) the measurements were made more often.
(c) the conditions were checked during the experiment.

> **The student has not made direct reference to the original *hypothesis*; is it accepted or rejected?**

> **Good point**

> **Possibly, though the data is fairly uniform, so the result might simply be more data points on the graph lines**

> **A good point, as this would probably affect growth rate. However, this could be explained more fully: does 'conditions' mean temperature, humidity, etc?**

Possible skills assessed in this coursework

'Experimental design and problem solving' (Ch 3, Units 2–5)

Examiner assessment

Moderate standard. The original hypothesis was stated in a rather general way, and this makes an evaluation of the investigation difficult. This is perhaps why the student has not really evaluated the investigation! Also, although the student has obviously 'thought out' the investigation, the plan has not been included, and the raw data have also been omitted. However, the student has summarised and presented the data very clearly. The conclusion and suggestions are good, though they could be developed.

Suggestions for further work

This experiment: Measurements continued for a longer period. Fresh or dry mass of a sample determined (e.g. two plants/week).
Related work: Growth rates in different conditions, e.g. temperature, light intensity, density of planting.
Linked work: Development (Unit 10), Variation (Unit 12), Abiotic (climatic) factors (Unit 18), Populations (Unit 19).

UNIT 10 GROWTH AND DEVELOPMENT – DEVELOPMENT

Background

Development is a process which normally accompanies *growth*. Development involves an increase in complexity of an organism; its cells and tissues become differentiated for specialist purposes. Development is a qualitative process, and changes due to development are often *described*. (This is different from growth, which is a quantitative process, and can be measured.) Development can be associated with fairly dramatic changes in body form ('morphology') in plants and, especially, in animals.

This sample coursework involves differences in development in two germinated seedlings (an example is shown in Fig 4.9). The student is told that the two seedlings are of the same species and were germinated at the same time. The practical involves (a) making suitable drawings of the two specimens, and (b) commenting on any differences between them. *R.G. X-ref.:* p 67, pp 72–73.

> **This exercise is a good example of a practical involving *observation* skills**

Fig 4.9
Example of germinated pea seedlings: *Left*, etiolated (grown in the dark) and *Right*, non-etiolated (grown in the light)

Title: Comparison of two germinated seedlings

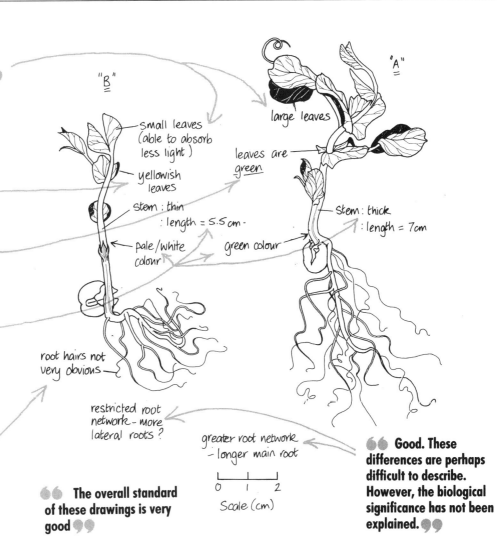

> **A good feature (in fact, light may not have been available for plant 'B' in any case). Note: leaves could be *measured***

> **A very important feature. Light is required for the development of chlorophyll; this suggests that plant 'B' did not have access to light**

> **Good observation. (The significance of this is explained above)**

> **These are important observations. These observations have not been explained**

> **Any difference in the seeds? These could be cut open to reveal any differences inside**

> **These are not visible on the drawings; additional, high-detailed drawings could show them**

> **The overall standard of these drawings is very good**

> **Good. These differences are perhaps difficult to describe. However, the biological significance has not been explained.**

Possible skills assessed in this coursework

'Observing and measuring' (Ch. 2, Unit 4), 'Recording and communication' (Ch. 2, Unit 5), 'Interpreting information' (Ch. 2, Unit 6)

Examiner assessment

Moderate-High standard. The quality of the drawing is excellent and shows very good observation skills. The interpretation was fairly good, though limited; the student has not shown an awareness that plant 'A' was germinated in the light, plant 'B' was germinated in the dark (i.e. it is 'etiolated'). Further direct comparisons between 'A' and 'B' were possible using more quantitative work (i.e. measuring, counting). This would have allowed more *relative* comparisons between stems, leaves and roots.

Suggestions for further work

This experiment: Conducting food tests on different regions of the germinated seedlings.
Related work: Experiments to show the different conditions needed for germination, e.g. moisture, warmth, oxygen, light, etc. Experiments to demonstrate respiration in germinating seeds, e.g. oxygen absorbed, carbon dioxide and heat evolved.
Linked work: Growth (Unit 9), Food analysis (Unit 17)

UNIT 11 HEREDITY AND VARIATION – HUMAN HEREDITY

Background

> **Genetics can be a difficult area in Biology. A good theoretical understanding will help in interpreting practical results**

Heredity is the study of the way in which different characteristics are transferred from one generation to the next. These characteristics are determined by *genes,* carried within the *chromosomes* of sex cells (gametes) of the parents. Each characteristic in the offspring is determined by one or more gene. Each gene can be expressed in two alternative forms, or *alleles.* In the *simplest* patterns of inheritance:

▸ each characteristic is determined by a single gene, and
▸ each allele of the gene is either *dominant* or *recessive.* Any dominant allele which is present in the chromosomes will be expressed. Recessive alleles are only expressed if no dominant allele is present.

It is difficult to study human genetics in practical Biology at GCSE. This is partly because relatively few of the obvious human characteristics are controlled by a single gene. One exception, however, is tongue-rolling (Fig 4.10). Individuals can either roll their tongue lengthwise, or not. The ability to roll the tongue is determined by the presence of one (or two) dominant alleles. An inability to roll the tongue is determined by the presence of a pair of recessive alleles. In this example of coursework, a survey of tongue-rolling ability was conducted within a class of pupils and their parents. *R.G. X-ref.* pp 81–86, 87–89.

Fig 4.10
Tongue rolling (right) and non-rolling (left)

CHAPTER 4 HEREDITY AND VARIATION – HUMAN HEREDITY

> ❝ This table of 'raw results' is difficult to interpret. It is possible to present the information in a more organised way, for example by giving the total numbers of students in each of six sub-groups; R students with parents both R, R students with one parent R, R students with both parents r, r students with both parents r, etc. (Note that some of these groups will be 'empty'). ❞

> ❝ No real use is made of this information, unfortunately! The pupil should ideally (a) ensure that information is obtained for a known purpose, or at least (b) make some comment about it. In fact, this information does show an interesting trend, and this could have been referred to! ❞

> ❝ A good graph; a bar chart is a very appropriate means of displaying this sort of information. ❞

> ❝ A good observation, though the numbers refer directly to the two groups, e.g. 'More pupils (14) were tongue-rollers than non tongue-rollers (5)'. The relative numbers of tongue-rollers: non tongue-rollers could be shown as a ratio, i.e. 2.8:1. ❞

> ❝ This is a good point, though 'dominant' has a special meaning in genetics, and this needs to be explained carefully, for instance by referring to genes or alleles. ❞

> ❝ This statement could be written more clearly. ❞

Title: Survey of tongue-rolling ability

RESULTS OF SURVEY

symbols used

R = tongue roller ; r = non-tongue roller

Student number	tongue rolling ability	PARENTS Father	mother
1	R	R	r
2	R	r	R
3	r	r	r
4	R	R	R
5	R	R	r
6	R	R	R
7	r	r	r
8	R	r	R
9	R	r	r
10	R	r	R
11	R	R	R
12	R	?	?
13	R	R	R
14	R	R	R
15	r	r	r
16	R	R	R
17	R	R	?
18	r	r	r
19	R	R	R

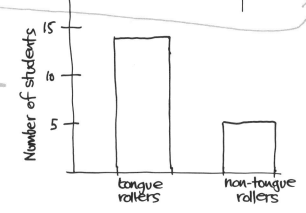

CONCLUSION

More students were tongue-rollers than non-tongue-rollers. The numbers were 14 and 5. This shows that tongue-rolling is dominant to non tongue-rolling.

CRITICISMS

1. It was sometimes difficult to get all the information we needed.

2. The number of students in the survey could have been more.

> ❝ An important point; a larger survey would provide more reliable information. ❞

Possible skills assessed in this coursework

'Recording and communication' (Ch. 2, Unit 5), 'Interpreting information' (Ch. 2, Unit 6).

Examiner assessment

Low standard. Conducting the survey may seem simple enough, but interpreting the information requires a fairly good understanding of genetics and the student has not demonstrated this. The task of interpretation is made more difficult in this case by an incomplete use of the information obtained.

Suggestions for further work

This experiment: Drawing of 'genetics diagrams' to illustrate the patterns of inheritance for this gene. Extending the survey to include brother/sisters, or even grandparents, for a limited number of pupils.
Related experiments: single-gene inheritance involving the presence/absence of distinct ear lobes, or the ability to taste PTC (phenylthiocarbamide).
Linked work: Variation (Unit 12).

UNIT 12 — HEREDITY AND VARIATION – VARIATION

Background

66 **Practicals involving variation provide a good chance to use data-handling and -presentation skills** 99

Even very closely-related organisms often show considerable variation, for instance in obvious characteristics such as size and shape. Variation results from a combination of 'internal' and 'external' factors, such as heredity, age and nutrition. There are two main types of variation:

- *continuous variation* This often results from several interacting factors. Organisms show a wide variety of different forms of characteristic (an example is shown below).
- *discontinuous variation* This often results from a single influence. Organisms show a very limited range of different forms of a characteristic, with few, if any, intermediate forms (see Unit 11 for an example of this).

Coursework involving continuous variation is common in GCSE Biology. In the example below, the student was asked to collect twenty privet leaves, measure the length of each one, present the results in a suitable way and comment on the results. *RG X-ref.* p 81, pp 89–90.

Fig 4.11
Measuring the length of privet leaves

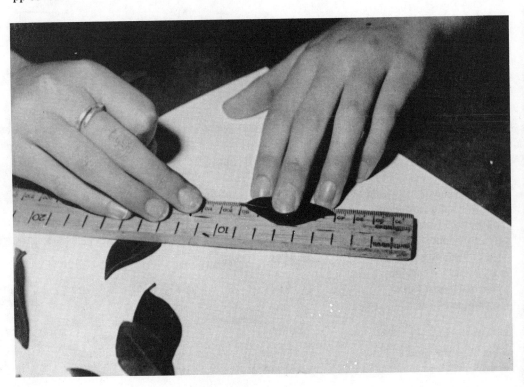

CHAPTER 4 HEREDITY AND VARIATION – VARIATION

Title: Variation of leaf length in privet leaves

RESULTS

RAW DATA

40, 43, 41, 12, 30, 19, 23, 45, 16, 51, 30, 50, 42, 37, 21, 21, 29, 39, 39, 49,

> Good choice of size groups

FREQUENCY TABLE

Size group (mm)	Number of leaves	
10 – 19	III	= 3
20 – 29	IIII	= 4
30 – 39	HHT	= 5
40 – 49	HHT I	= 6
50 – 51	II	= 2
		20

> Giving the total is a useful way of checking that all leaves have been included.

> A clear, well-organised graph

> 'Size' is too vague here; 'length' would be better

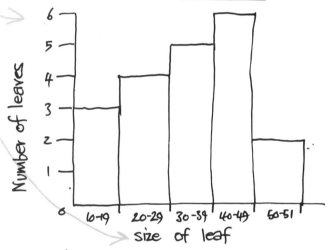

> An excellent summary of the data, giving a clear idea of the distribution of data

CONCLUSION

The length of privet leaves varied between 12mm (minimum) and 51mm (maximum). The mean length was 677 ÷ 20 = 33.9 mm. Most leaves had a length in the range 30 – 49 mm.

> What does 'most' mean? There are 11 leaves in this range.

There are several possible reasons why there was a variation in length amongst the privet leaves:

1. The leaves were of different ages; older leaves are usually longer than younger leaves.
2. The leaves were from different plants.
3. The leaves received different amounts of light so did not all photosynthesize at the same rate. This would affect the amount of growth.
4. The leaves received different amounts of water and minerals.
5. Some leaves shaded the light from other leaves, limiting the growth of the leaves below.

> A good point. Some general reference could be made to the relative *position* of different-sized leaves on branches, or on the plant.

> Yes. This would perhaps account for any genetically-determined variation amongst the leaves.

> An important point. Did the student make observations of light levels whilst collecting the leaves? Were leaves collected from both North- and South-facing sides of the privet hedge?

> A valid point (though evidence for this would be rather difficult to obtain)

CRITICISMS

1. The leaves might not have been chosen randomly, for instance the very small leaves may have been ignored.
2. The measurements may not have been accurate.

> Good. Again, is this supported by observations? The effect of shading could account for length variation. However, shading could increase or decrease leaf length, depending on the species.

> Good point. Some suggestions for avoiding this might be made here.

> Yes; this might apply in particular to the smaller leaves because they can be difficult to measure.

Possible skills assessed in this coursework

'Observing and measuring' (Ch 2, Unit 4), 'Recording and communication', (Ch 2, Unit 5), 'Interpreting information' (Ch 2, Unit 6).

Examiner assessment

High standard of work throughout. The information has been clearly and economically organised, and the interpretation of results has been fairly thorough. However, the student could have made more reference to the particular site where the leaves were collected. The results show a 'skew' rather than a symmetrical ('normal') distribution; the mean value is outside the 'mode' group containing the most leaves. This *may* mean that the sample was not very representative.

Suggestions for further work

This experiment: Increasing sample size, for instance by combining class results. Investigating possible environmental and other differences at the site where the leaves were collected.
Related experiments: Measuring variations in mass and length of different species. Measuring differences in humans, e.g. height, index finger length. Measuring variations in petal number in daisy flowers.
Linked work: Diversity of life (Units 5 and 6), Growth (Unit 9), Climatic factors (Unit 18).

UNIT 13 — RESPIRATION – INTERNAL RESPIRATION: ANAEROBIC

Background

Respiration is the process by which energy is released within living tissues; it is a characteristic of all living things. For convenience, respiration can be divided into two distinct phases:

- *External respiration* This consists mainly of *physical* processes, involving the exchange of gases between the organism and its environment.
- *Internal respiration* This consists mainly of *chemical* processes, involving the release of energy during the breakdown of relatively large molecules to smaller molecules. If this breakdown involves oxygen it is called *aerobic respiration*. If oxygen is not involved, the breakdown is called *anaerobic respiration*.

> 66 **Some understanding of the chemical reaction involved would be useful in this experiment!** 99

This example of coursework is based on anaerobic respiration in yeast. The yeast releases energy by breaking down its food (glucose) to the waste products carbon dioxide and alcohol. This is an enzyme-controlled reaction and is therefore affected by temperature. The rate of this reaction can be followed by observing the number of bubbles of carbon dioxide released in a given time (Fig 4.12). *RG X-ref:* pp 107–109, p 111–112.

Fig 4.12
Anaerobic respiration in yeast

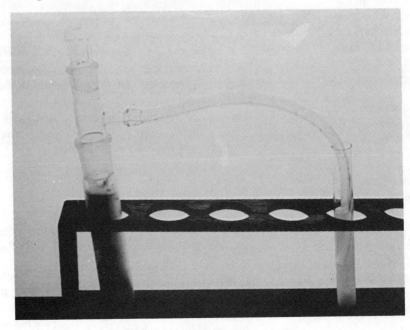

CHAPTER 4 RESPIRATION – INTERNAL RESPIRATION: ANAEROBIC

Title: The effect of temperature on the respiration of yeast

> **A good, well-presented table of results.**

Results

time (mins)	Number of bubbles per min.	
	30°C	40°C
1	0	10
2	0	10
3	0	9
4	0	9
5	1	9
6	2	8
7	2	7
8	2	6
9	3	5
10	3	4
mean value	1.3	7.7

> **An excellent explanation of these results; the student shows a good understanding of the theory.**

> **Possible but unlikely to have much effect within 2–4 minutes**

> **A very good explanation.**

> **What is the significance of this? Is the rate of multiplication more rapid at 40°C?**

> **It is best to avoid giving an interpretation as being 'due to' certain 'facts'! Another way of putting this might be: 'This may be caused in various ways, including:' etc.**

> **More accurate to say 'alcohol inhibiting the enzyme activity'.**

> **Yes, though this is a gradual process.**

> **The effect on the *enzymes* involved in the reaction is more relevant here. 40°C is not usually regarded as 'excess heat' for enzyme reactions; in fact, it is close to the optimum temperature for most enzymes.**

Conclusions

At 30°C the reaction is very slow. For the first 4 minutes no bubbles are produced. This is because the enzyme needs to be activated. It takes some time for the heat of the surrounding liquid to conduct through the glass boiling tube into the solution. Also the enzyme takes some time to produce the correct enzymes to break down the glucose. The yeast begins to reproduce. This causes the rate of reaction between the glucose and the yeast to increase, because the molecules come into contact more often because they have more energy.

At 40°C the rate of reaction increases due to the greater amount of energy. Therefore in the first 2 minutes a lot of CO_2 is produced. Also, the yeast is multiplying.

After 2 minutes at 40°C the reaction time is decreasing.

This is due to the following facts:
i) Alcohol is killing off the enzyme
ii) Sugar is all used up
iii) Yeast may become denatured by the excess heat

Possible skills assessed in this coursework

'Following instructions' (Ch. 2, Unit 2), 'Observing and measuring' (Ch. 2, Unit 4), 'Interpreting information' (Ch. 2, Unit 6).

Examiner assessment

High to moderate standard of work. The presentation of the results is good; the student should also have drawn a graph (like the one shown here) to provide a more immediate comparison of the situations. The conclusions revealed a very good understanding of many of the processes involved. However, more direct reference could have been made to the data. For instance, was the rate of reaction constant for each temperature? What do the mean values show? Was the temperature of both reaction mixtures maintained throughout?

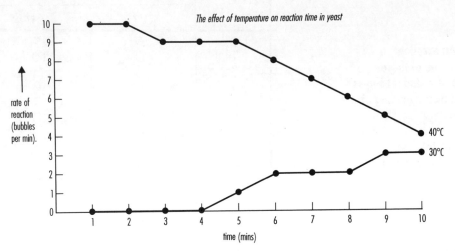

Fig 4.12b
Graph based on the pupil's results

Comments on the accuracy and reliability of an experiment are important. They can also be used to make suggestions for improvements. Experiments need not be 'perfect' (there is probably no such thing!), but students can demonstrate their understanding by pointing out limitations in the techniques used and in some of the assumptions made.

Suggestions for further work

This experiment: Repeated measurements, including a wider range of temperatures. Conducting the experiment with sucrose rather than glucose.
Related experiments: Collecting and testing the gas given off by yeast during anaerobic respiration. Measuring the expansion of dough (in a measuring cylinder) at different temperatures. Measure the production of alcohol (using a hydrometer) in the reaction mixture.
Linked work: Enzymes (Unit 2).

UNIT 14 TRANSPORT SYSTEMS – BLOOD SYSTEMS: THE HEART

Background

66 This exercise is *very* popular amongst teachers for coursework assessment! 99

The purpose of transport systems is to move substances from regions where they are made or where they enter the organism to regions where they are used or where they leave the organism. This process is necessary because different parts of organisms are often specialised for different functions.

In many plants and animals there is a *vascular system* consisting of a network of tubes carrying various fluids. The vascular system in larger, more active animals may include a pump to circulate the *circulatory fluid*. In mammals such as humans, the main vascular system consists of blood vessels and includes a heart to circulate blood. The rate at which the heart operates varies according to the body's need for the substances which are supplied or removed by the blood. Examples of such substances include oxygen, food and carbon dioxide.

In this example of coursework, the student was given a hypothesis by the teacher and asked to design and perform an appropriate experiment. The student decided to measure the pulse rate of a subject (Fig 4.13) at intervals following a period of intensive exercise. *RG X-ref.* pp 139–40, p 146, p 155.

Fig 4.13
Measuring pulse rate

Title: Experiment to show the effect of exercise on heart rate

❝ This hypothesis, suggested by the teacher, should be in the basis of the investigation. ❞

❝ Where is the plan for this investigation? Obviously, the student *has* planned the experiment; this may have been discussed with the teacher, who can make an assessment as well. However, it is also a good idea to give a written outline of the plan, too. This may be needed for the examiner to assess later. ❞

❝ How was this done? Some (brief) details could be included here. ❞

❝ The pulse should really be taken *before* exercise, too. In practice, taking the pulse *during* exercise is usually difficult, however. ❞

❝ *Units* should be given for both columns! ❞

HYPOTHESIS: That the pulse rate returns to normal more quickly in a fit person compared with an unfit person.

METHOD The pulse of a volunteer who does a lot of sport after school was taken after excercise. The exercise was to step on and off a block for 2 minutes. The pulse was taken immediately after the excercise and continued for 10 minutes.

time	pulse rate
0	97
1	51
2	40
3	39
4	35
5	35
6	34
7	35
8	34
9	35
10	34

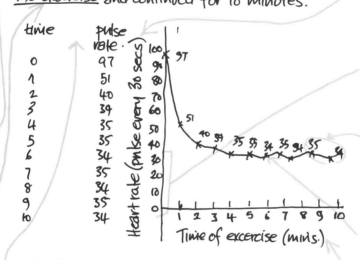

CONCLUSION

The results show a rapid drop in the pulse rate after exercise. After 4 minutes the pulse rate had reached its normal level. The most rapid drop occured in the first minute.

The experiment shows that a fit person's heart rate had a good recovery rate compared with an unfit person, where it took longer for the normal level to be reached.

❝ Although it is easier to *measure* pulse over 30 seconds rather than 1 minute, this should be doubled when *presenting* the data as a graph. This graph is rather confusing in this form. A much-improved scale would be 'pulse counts per minute'. ❞

❝ This part of the scale is not strictly necessary. It could start at, say, 60 pulse counts *per minute*. ❞

❝ Good. Writing the actual values onto the graph can be useful provided they do not obstruct the graph itself. ❞

❝ This point is incorrectly plotted. ❞

❝ A generally useful graph, although the y-scale is misleading. ❞

❝ A very good summary of the results. Note that continuing the readings is not really necessary after about 5 minutes! ❞

❝ Where are the results for this? They are essential if the student intends to prove or disprove the original hypothesis! ❞

CHAPTER 4 STUDENT COURSEWORK WITH EXAMINER COMMENTS

Possible skills assessed in this coursework

'Experimental design and problem-solving' (Chapter 3)

Examiner assessment

Low standard. The student has shown a fairly good ability to record results, but the interpretation is quite limited. For instance, little detailed reference is made to the actual results, including the changing heart rate after exercise. The student should have included the data from the 'unfit' subject, and a reference to the hypothesis made. If the student had adopted a different hypothesis, this should have been given. The student could have gained more credit by giving *criticisms* of the investigation and *suggestions* for further investigation.

Suggestions for further work

This experiment: Comparing the effect of exercise on pulse rate of individuals from different groups, e.g. age, sex, smoking/non-smoking, on pulse rate. Determining the effects of caffeine on pulse rate.
Related experiments: Comparisons of changes in the rate of breathing and pulse rates.
Linked work: Variation (Unit 12).

UNIT 15 SENSITIVITY – ANIMAL SENSITIVITY: THE SKIN

Background

> **There are numerous other practicals on sensitivity, and also behaviour**

Sensitivity is a characteristic of life. All organisms are capable of detecting and responding to changes in their environment. A change in the environment is a *stimulus* which, in 'higher' organisms, is detected by specialised *receptor* cells, tissues and organs. In humans, a good example of a sense organ (which also performs other functions!) is the skin.

In this example of coursework, the student has written up a class experiment on skin sensitivity. The experiment is designed to compare the sensitivity to touch of different parts of the body surface (Fig 4.14). *RG X-ref:* pp 159–161, pp 164–168

Fig 4.14
Testing skin sensitivity using two mounted pins

Title: Comparison of skin sensitivity of different parts of the body

> **Good. The method has been clearly and briefly explained.**

METHOD

One or two pin heads were placed lightly against the skin at three positions on the body. The subject did not look at the pins when the test was made. The subject was asked to say whether one or two pins were in contact with their skin; the number of correct responses out of ten trials was recorded

The distance between the pins was also varied.

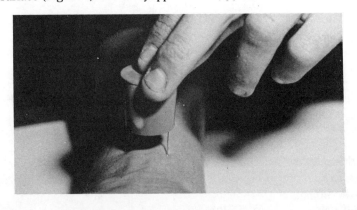

The results for the whole class were combined and mean values were calculated. A graph was drawn of the results

> **A good summary of the class results. However, *raw data* should also be included so that calculations can be checked by the teacher or examiner.**

RESULTS	DISTANCE BETWEEN PINS (CM)		
POSITION ON BODY	2.0	1.0	0.5
PALM	9.2	8.1	5.9
BACK OF HAND	7.4	4.4	5.2
BACK OF NECK	6.6	5.4	5.4

TABLE SHOWING MEAN NUMBERS OF CORRECT RESPONSES OUT OF TEN TRIALS

> **A good, clear graph. The block for each pin setting (e.g. 2.0, 1.0 etc.) should be shaded differently. This would make comparisons between the groups easier.**

> **These numbers should be explained, e.g. in key accompanying the graph.**

> **Good; well-presented.**

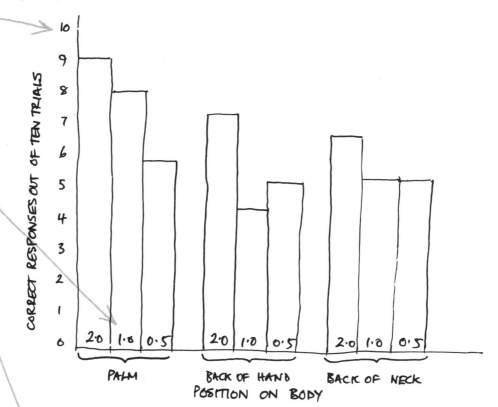

> **The sensitivity of the skin is not actually affected by the position of the pins! However, the ability to correctly identify the number of pin heads *is* affected.**

> **Excellent. This shows a very good understanding. However, this is a complex idea, so the student could provide *sketch diagrams* to illustrate the interpretation.**

> **This is not shown in this account because raw data has not been included.**

> **It is important to be able to identify 'unusual' results. However, some explanation is also needed.**

CONCLUSION

The sensitivity of the palm was higher than the other two regions for each pin setting. When the pins were set at 2cm, the order of decreasing sensitivity was:

Palm ⟶ back of hand ⟶ back of neck

The sensitivity of the back of hand and ~~and~~ neck were similar when the pins were close together.

Reducing the distance between the pins decreased the sensitivity of the skin in all cases. This is because there is a reduced possibility of two nerve endings being stimulated together, or one nerve ending being stimulated.

CRITICISMS

This experiment seemed to work well though there was quite a variation in the class results for each situation. The result for 1cm with the back of the hand seems quite low.

IMPROVEMENTS

1. More positions on the body could be tried.
2. More distances between the pins could be used.

Examiner assessment

High standard. The student has presented the results sensibly, and has demonstrated a very good understanding of the processes involved.

Suggestions for further work

This exercise: Extend the range of positions tested on the skin surface. Extend the range of different pin settings. Use a grid to systematically test a small area of skin.
Related exercises: Test skin receptors for pressure, heat and cold. Place each hand separately in hot or cold water, then transfer them together into tepid water; report on the sensation. Map the surface of the tongue for different sweet, sour, salt and bitter tastes.

UNIT 16 — NUTRITION – AUTOTROPHIC NUTRITION: RATE OF PHOTOSYNTHESIS

Background

Nutrition is the process by which all living organisms obtain materials from their environment for energy and growth. There are two main types of nutrition:

- *Autotrophic nutrition* or *photosynthesis*. This is used by green plants, and involves building up fairly large, complex molecules from much simpler molecules absorbed from the environment. The energy for this process comes from the sun, and is absorbed by the green pigment *chlorophyll*, mainly contained in the leaves. Also absorbed for this proccess is carbon dioxide and, in bright light, the waste gas *oxygen*, is released by the plant.

- *Heterotrophic nutrition*. This is used by animals (and also any other organisms not capable of photosynthesis). The process mainly involves taking in large molecules from the environment, often by consuming other organisms, and breaking them down to smaller molecules. These can then be used to release energy (see Unit 17) or to build new molecules.

❝ **This experiment provides a good opportunity to demonstrate skills in *handling apparatus and materials*** ❞

The experiment described in this example of coursework will be familiar to many students taking GCSE Biology. The effect of different light intensities on photosynthesis is studied, using an aquatic plant called *Elodea* (Canadian pondweed). The advantage of using an underwater plant is that bubbles of oxygen can be clearly seen. The rate at which they are released can be counted (Fig 4.15), and this is assumed to correspond with the rate of photosynthesis. *RG X-ref* pp 187–189, pp 193–194.

Fig 4.15
Measuring the rate of photosynthesis in Elodea (distance between lamp and beaker is 20 cm).

Title: Experiment to show the effect of different light intensities on the rate of photosynthesis

METHOD

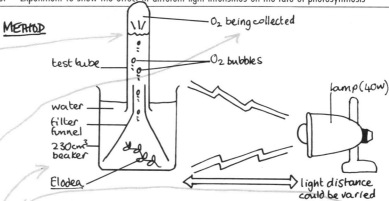

> The diagram should be planned properly *before* it is drawn; it is a good idea to start with the top of the diagram (in this case, the test tube) and work down. Better still, plan it in light pencil first.

> Note: the gas bubbles are only *assumed* to be oxygen, this can be confirmed if sufficient gas is collected for testing.

> Teachers and examiners prefer pupils not to draw 'cartoon-style' diagrams like this; it loses accuracy and is distracting. The student could have made the diagram less 'crowded' by labelling only the really important features.

1. A piece of Elodea was cut from a plant underwater and quickly transferred to a 250 cm³ beaker (see diagram). A paper clip was attached to the plant to keep it down under water. A 40w light was placed next to the beaker: distance = 0cm

2. Next, the number of (large) bubbles released by the plant during one minute was counted. This was repeated three times. The mean number of bubbles ~~was counted~~ released per minute was calculated.

3. The distance between the beaker and the lamp was then increased to 5 cm. The mean number of bubbles released per minute was again calculated.

4. This procedure was repeated at the following distances: 10cm, 20cm, 30cm.

> Numbering the different stages of the method like this is a good idea if there are several steps involved.

> Good; details like this are worth including.

> The importance of this could be briefly explained.

> A good description.

> Twice, to be exact.

RESULTS

DISTANCE (cm)	SEPARATE MINUTE INTERVALS			MEAN VALUES
	①	②	③	
0	22	35	38	31.666
5	36	31	30	32.333
10	22	18	17	19
20	6	5	7	16
30	2	4	2	2.666

> There should be some reference to what is actually being measured, i.e. 'number of bubbles released per minute'

> There is no need to express calculated values to several decimal places. Also, these values imply a level of accuracy not actually achieved in the measurements.

A GRAPH BASED ON THE TABLE OF RESULTS.

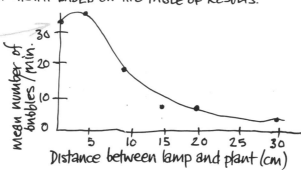

> A very good graph. The student has avoided a common error of drawing a histogram for this sort of data.

> Very good summary of the results. The student has made the important links between *distance* and *light intensity*, and between the *number of bubbles released* and *photosynthesis*.

CONCLUSION

As the distance between the plant and the lamp increased, the light intensity decreased and the mean number of bubbles (or rate of photosynthesis) decreased. This is clearly shown by the graph.

> **A good point.** There is in fact evidence for this in the data; the number of bubbles given off increases with time during the first set of readings. This can be avoided by (a) delaying the first readings for about 5 minutes, and (b) putting a sheet of glass between the beaker and lamp, to absorb some heat.

CRITICISMS
1. The heat from the lamp could have affected the plant.
2. The results varied a lot, at each distance. This shows the importance of mean values.
3. Light from other experiments or the room lights might have affected the results.

> **An excellent point.** Combining class results can be useful, if experimental conditions are the same.

> **Good point.**

Possible skills assessed in this coursework

'Following instructions' (Ch. 2, Unit 2), 'Handling apparatus and materials' (Ch. 2, Unit 3), 'Recording and communication' (Ch. 2, Unit 5), 'Interpreting information' (Ch. 2, Unit 6)

Examiner assessment

Moderate standard of work. The student has obviously conducted the experiment fairly well, though could have improved the reliability of the results by a more careful approach. The results (especially the graph) are presented well, though the student omits any observations other than the number of bubbles. For instance, was bubble size variable? Were the bubbles released at regular intervals? Where on the plant were the bubbles released from? Was there any noticeable change in the temperature of the water during the experiment? (Ask for a thermometer, or feel the outside of the beaker).

The interpretation is concise and relevant, though rather general. The student might have made more *detailed comments* on the actual data and graph, and also the limitations of the experiment.

Suggestions for further work

This experiment: Vary light intensity by using different wattage light bulbs, e.g. 60 W, 100 W (distance constant). Test the effect of changing temperature at a constant light intensity. (CARE! **Do not pour water near a bench lamp. Do not touch the lamp with wet fingers.**) Raise the CO_2 content of the water by adding (known?) quantities of sodium hydrogen carbonate (sodium bicarbonate) powder. Collect and test the gas (using a glowing splint). Measure the volume of gas collected (using an inverted 10 cm^3 measuring cylinder).
Related experiments: Observe colour changes in *Elodea* in bicarbonate indicator (changes colour with changing CO_2).
Linked work: Internal respiration (Unit 13), Climatic factors (Unit 18).

UNIT 17 NUTRITION – HETEROTROPHIC NUTRITION: FOOD ANALYSIS

Background

> **This exercise is one of the most commonly used in coursework assessments**

Nutrition involves organisms obtaining materials from the environment for use in essential processes. The two main purposes of food molecules are:

- *Growth and development.* This often invovles building up larger molecules from smaller molecules. For example, building up protein molecules from amino acids, or (in plants) cellulose from sugars.
- *Release of energy.* This involves breaking larger molecules into smaller molecules, in *respiration*. For example, breaking down glucose into water, carbon dioxide and energy, using oxygen; this is called *aerobic respiration*. Food represents stored energy, and this energy can be released within living tissues by respiration, or outside the body by *burning*. The energy value of food depends on the amount of food present (i.e. its mass) and also what types of molecules it contains (its composition). The relative energy value of different foods varies because they contain different proportions of various molecules.

The experiment shown in this coursework occurs *very* commonly in coursework assessments! It is a fairly easy experiment to perform, but there is much scope for the interpretation of results and, in particular, criticisms of the method. The experiment involves burning a known mass of food and using the heat energy released to raise the temperature of some water (Fig 4.16). *RG X-ref.* pp 197–198, pp 200–201.

Fig 4.16
Releasing the energy from a peanut

Title: Comparison of the energy content of three foods

> More information could have been presented here. Marks are not always awarded for writing up the 'method' section, but it is important (i) in experiments designed by the student (marks given for plan) and (ii) interpretation of information; assumptions, limitations, etc. will need to be referred to.

METHOD
The experiment involved burning three different foods. The energy released from this was used to heat 20 cm³ water in a boiling tube, and the temperature increase of the water was measured. The energy value of each food was calculated using the formula below:

$$\text{ENERGY VALUE OF FOOD (J)} = \frac{\Delta t \times S \times m_w}{m_f}$$

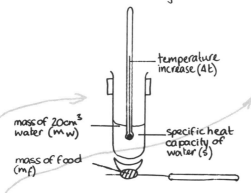

KEY Δt = temperature increase (°C)
S = specific heat capacity of water = 4.2 J/g
m_w = mass of water (g)
m_f = mass of food (g)

> A useful diagram, showing the method and also summarising the measurements made.

RESULTS

FOOD SAMPLE	PEANUT	APPLE	TOAST
m_f Mass of food (g)	0.6	2.3	0.4
m_w Mass of water (g)	20	20	20
Initial temperature of water (°C)	18	19	19
Final temperature of water (°C)	54	23	25
Δt Temperature increase (°C)	36	4	6
Energy value of food (J)	$\frac{36 \times 4.2 \times 20}{0.6} = 5040$	$\frac{4 \times 4.2 \times 20}{2.3} = 146$	$\frac{6 \times 4.2 \times 20}{0.4} = 1260$

> Good. It is important to show how calculations have been done. If symbols are used, their meaning must be given; the pupil has done this.

> A very clear, informative table.

> A good summary of the results.

CONCLUSION
The peanut contained the most amounts of energy per gram (5040 J/g, or 5.04 kJ/g.), the apple contained the least amount (146 J/g). The toast had an intermediate value (1260 J/g = 1.26 kJ/g).

> Differences in energy value may occur for a variety of reasons, including experimental methods and in the calculations.

Differences in energy value between the three foods was due to the different sorts of molecules that they contain. For instance fats release a relatively large amount of energy, carbohydrates and proteins rather less. Water does not release any energy. Energy is released in living cells by respiration which requires oxygen and which releases carbon dioxide. The release of energy from burning is therefore similar to the release of energy during respiration.

> This is rather 'theoretical' and does not make close reference to the results of *this* experiment.

However, the amount of energy calculated in this experiment is likely to be less than that released if the food is eaten and then used in respiration. There are several ways in which heat energy is lost in this experiment:

❝ **Good point.** ❞

1. The food is not fully burned. It was quite difficult to get the food to burn, especially, the apple.

❝ **These seem to be covering the same general point, though it *is* a good point!** ❞

2. Heat may escape from the food when it is being transferred from the bunsen burner to the boiling tube, and when it is held under the tube, the flame does not just heat the water.

3. Heat may be conducted away from the water by the air, glass, mounted needle.

Possible skills assessed in this coursework

'Handling apparatus and materials' (Ch. 2, Unit 3), 'Observing and measuring' (Ch. 2, Unit 4), 'Interpreting information' (Ch. 2, Unit 6), 'Carrying out safe working procedures' (Ch. 2, Unit 7).

Examiner assessment

Moderate standard. The standard of this coursework is variable, however. The results and calculations have been very well presented; the interpretation (including the criticisms) are reasonable, though lacking in 'depth'. Also, the student could have included *observations*, e.g. of the colour, intensity of the flame, whether sooty deposits were left on the underside of the boiling tube and the appearance of the food after burning. The student has not referred to possible inaccuracies in the measurements. This is a fairly short experiment to perform and, if time and materials allow, the student could repeat all or part of the experiment. Another idea is to refer to the results of other groups; results can be 'pooled' and values compared. Mean values can be compared.

Suggestions for further work

This experiment: Different forms of each of these foods could be used, e.g. roasted/unroasted peanuts, bread/toast, fresh/dried apple. A wider range of food could be used, including currants, raisins etc.
Related work: Food can be fairly simply tested for starch, reducing sugars, protein, lipids, vitamin C (see Longman 'Revise Guide: GCSE Biology', p 200).
Linked work: Internal respiration (Unit 13).

▶ UNIT 18   ENVIRONMENT – ABIOTIC FACTORS: CLIMATIC FACTORS

Background

❝ **Note that the distribution of organisms is usually determined by *several* interacting factors!** ❞

Biotic factors are the living (biological) components of the environment. In other words, the biotic part of the environment involves organisms and their direct interactions with each other (Unit 19).

Abiotic factors are the non-living (physical and chemical) components of the environment. Examples of abiotic factors include the climate, the soil and circulating minerals. Each abiotic factor interacts with many others and, of course, with the biotic components. In fact, it is often difficult to separate the relative effects of living from non-living parts of the environment.

The coursework sample shown here is based on an exercise which is fairly commonly used in assessments. The simple alga *Pleurococcus* grows on the surface of many trees. It appears as a green 'dust' and its distribution seems to be affected by two related climatic influences, temperature and moisture. These abiotic factors vary according to the direction of the tree surface. A systematic way of estimating *Pleurococcus* distribution is shown in Figure 4.17. *RG X-ref.* pp 213–215.

Fig 4.17
Transparent quadrat used to measure *Pleurococcus* distribution

Title: Experiment to investigate the distribution of Pleurococcus on tree surfaces facing different directions

❝ How was this done? More details needed! ❞

Apparatus   compass   metre rule   a tree!
quadrat (10x10)   string

❝ What size grid was it? ❞

Method   We selected a tree at random and used a quadrat to find out how much Pleurococcus was growing on it. The quadrat was a transparent 10x10 grid, and we had to count how many squares (out of 100) contained Pleurococcus. If more than half a square was occupied, it was included, otherwise not: ✓ ✗

❝ Good. Diagrams like this are helpful. ❞

❝ The raw data should be included; this allows the teacher or examiner to check the accuracy of the graph. ❞

The quadrat was placed on the tree 1 metre from the ground and tied there with string. The quadrat was moved to eight different compass positions, starting from North.

❝ A good, clearly-presented graph. (A line graph would not have been appropriate here). ❞

Results The results of our Pleurococcus survey are shown in the graph:

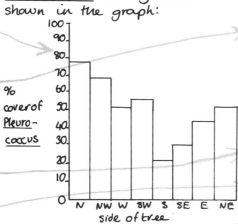

Conclusion There was more Pleurococcus growing on the North-facing side of the tree compared with the South-facing side. The percentage of Pleurococcus gradually changed as you went around the tree, and this is clearly shown in the graph. There are several reasons why there is a difference in the distribution of Pleurococcus:

❝ Good summary. ❞

❝ Good. There are *two* suggestions here; they could be presented separately. ❞

a) The North side might be cooler or moister.
b) The North side might be too bright.
c) Other plants or animals might be affecting it.

❝ Yes, possibly. ❞

❝ In what ways? More details needed here! ❞

Criticisms There are several things wrong with this experiment. For instance:

a) only one tree was used. Using more trees would give more certain results.
b) The quadrat may not have been placed in a position which showed what was really happening.
c) It was sometimes difficult to estimate the amount of Pleurococcus, eg. because of the uneven tree bark.

❝ Good point. Class results could be combined, and mean values used. ❞

❝ Good point. The 'smoothness' of the bark will vary between tree species. ❞

❝ This is not clear. 'Unrepresentative' results would be obtained by repeating the experiment (though *not* by carefully selecting where to put the quadrat!) Note that the age of the tree may be important, because of the time required for the *Pleurococcus* to colonise. Also, in younger trees with a narrower girth, the quadrats may overlap. ❞

Possible skills assessed in this coursework

'Observing and measuring' (Ch. 2, Unit 4), 'Recording and communication' (Ch. 2, Unit 5) 'Intepreting information' (Ch. 2, Unit 6)

Examiner assessment

Moderate standard. The student has presented the results well, though raw data have been omitted. The interpretation is quite good, though the pupil could have provided a more detailed explanation. For example, the biological significance of the observation could be discussed.

Suggestions for further work

This experiment: Comparison of *Pleurococcus* distribution on the surfaces of different tree species. Measurement of environmental factors, such as temperature, humidity, light. An assessment of the *density* of *Pleurococcus*, e.g. on a scale 0–3.
Related experiments: Comparison of the growth of nettle plants growing in shade and sun (see Chapter 3.) Comparison of the properties of different soil types (see Longman 'Revise Guide: GCSE Biology' pp 216–219).
Linked work: Natural classification (Unit 5), Populations (Unit 19).

UNIT 19 ENVIRONMENT – BIOTIC FACTORS: POPULATIONS

Background

> **Remember that *small* samples do not provide reliable information about the rest of the ecosystem**

Biotic factors consist of the living part of the environment. The relative abundance and distribution of different species is determined by *abiotic* (non-living) factors (see Unit 18 above) and by interactions between the organisms themselves. A major way in which oganisms interact is in *feeding relations*. Patterns and feeding interactions can be represented as food chains and food webs.

This example of coursework began as a class exercise, sampling the invertebrates of a stream (Fig 4.18). The coursework shown here is by a student who decided to study some of the feeding relationships of the animals in the class sample. *RG X-ref.* pp 232–235, pp 237–238, p 242.

Fig 4.18
Using nets to sample the animal life in a stream

Fig 4.19
Using a quadrat to determine the diversity and abundance of plant species (school field)

CHAPTER 4　ENVIRONMENT – BIOTIC FACTORS: POPULATIONS

Title: Survey of freshwater invertebrates in a stream

> **Excellent.** The information has been very clearly presented.

SPECIES	NUMBERS COLLECTED	PIE DIAGRAM ANGLE
Carnivores: Flatworm	77	
leech	4	$\frac{107}{245} \times 360$
caddisfly larvae	24	
diving beetle	2	$= 157°$
CARNIVORE TOTAL	= 107	
Herbivores: Snails	4	
Mayfly nymph	43	
halipid beetle	4	$\frac{55}{245} \times 360$
riffle beetle larvae	1	
riffle beetle	3	$= 81°$
HERBIVORE TOTAL	= 55	
Detritus feeders: freshwater worm	37	
freshwater shrimp	17	
water hoglouse	11	
water flea (ostracod)	1	$\frac{83}{245} \times 360$
water mite	3	
woodworm	13	$= 122°$
blackfly larvae	1	
DETRITUS FEEDERS TOTAL	= 83	
TOTAL OF ALL ANIMALS	= 245	

> **Good.** It is important to *show* calculations.

> **This information is important**, but there is no further reference to it in the conclusion!

Also caught: midge pupa (non feeding) 2
Mean spread of water flow = 0.28 meters per sec.
Mean oxygen concentration (% saturation) = 108.5%

PIE DIAGRAM OF THE DIFFERENT FEEDING GROUPS FROM THE STREAM SURVEY: FRESH WATER INVERTIBRATES

> **This pie diagram is rather small.**

> **Much of this information could be included in the results section. The percentage values could be included in the results table.**

CONCLUSION

The class results for the stream sampling, show that 16 different species of invertibrates were caught. The totals for each species are shown in the results table. I divided the species into three groups, according to their feeding habits. I have calculated the proportion of animals in each group:

carnivores = $\frac{107}{245} \times 100 = 44\%$

herbivores = $\frac{55}{245} \times 100 = 22\%$

Detritus feeders = $\frac{83}{245} \times 100 = 34\%$

> **Good.**

These percentage values and the diagram show that the largest group are the carnivores, then the detritus feeders, then the herbivores. This result can be shown as a simple food web: herbivores (55)
detritus feeders (83) → carnivores (107)

> **This is rather misleading!** The 'prey' animals included in the sample obviously have not yet been eaten! In fact, it cannot be assumed that they will necessarily be eaten by these 107 carnivores (or perhaps by any other carnivores). However, the *general* point being made is a fair one!

The total number of animals being eaten by the carnivores is 55 + 83 = 138. This is more than the number of carnivores in the total sample (107). A pyramid numbers for this survey would look like this: 107 carnivores
herbivores + detritus feeders   138

CRITICISMS

The experiment seemed to work quite well, but sampling may not have been very accurate. Here are some reasons for this:

1. We mainly sampled at the side and at the bottom of the stream, not in the central part. Some species may have been hiding under stones, etc.
2. The mesh of the nets may have let smaller animals through.
3. It was sometimes difficult to identify species and to count numbers. We may made mistakes.
4. Vertebrates (eg. Fish, tadpoles) were not included in our results. They may have been important in the food web.

> **Diagrams like this are a good idea.** However, the pupil should really use *proportions* rather than actual numbers. In this case, the relative numbers of potential prey and predator animals in this sample is 138:107, or 1:0.8.

CHAPTER 4 STUDENT COURSEWORK WITH EXAMINER COMMENTS

> **A simple yet highly-effective chart showing the results at a glance**

RESULTS

SPECIES	1	2	3	4	5	6	7	8	9	10	11	12	13	14	15	16	17	18	19	20	21	22	23	24	25	26	27	28	29	30
Cross-leaved heath											■	■	■	■	■	■	■	■												
Ling heather	■						■	■	■	■										■	■	■	■	■	■	■	■	■	■	■
Purple moor grass	■	■	■				■	■	■	■	■		■	■		■		■	■	■	■			■	■	■	■	■	■	■
Deer grass					■	■																								
Sphagnum	■	■	■			■					■	■	■	■	■	■	■	■	■	■	■	■	■	■	■	■	■	■		
Bracken																									■					

KEY: ■ SPECIES PRESENT   ☐ SPECIES ABSENT

CONCLUSION

The results show that different plant species occur at different positions along a transect. None of the plants occurred in every quadrat along the whole transect. There are several reasons for this:

1. Wetness

Plants prefer either wet or dry conditions. We used books to find out what these plants preferred:

'Wet-loving species': cross-leaved heath, purple moor grass, deer grass, sphagnum
'Dry-loving species': Ling heather, bracken

> **Things are not always this simple in biology! Some plants may not have a clear preference. Others may have a distinct preference, but they are also affected by many other factors**

> **Good, though it may be oxygen or pH levels or temperature that are more important than 'water' as such**

The results show that quadrats 2-5 and also 29 had wet conditions, as no 'dry loving' species were found there.

There was no overlap between the 'wet loving' cross-leaved heath and the 'dry loving' ling heather. This seems to indicate a particularly wet area in the centre of the transect (quadrats 11-18)

> **Good; well-explained**

> **Excellent**

In several quadrats, both 'wet-loving' and dry-loving species were found. This may be due to mixed conditions, or perhaps the plants being tolerant.

> **This shows a very good understanding of the principles involved**

2. Competition

As there is not a very clear 'wetness' effect on the distribution of all plant species, this means that other factors were involved. For instance, one type of plant might have been preventing another one growing by competition. This seems to be what is happening with the two types of heather, which we did not find together at all. Bell heather only grows in places where ling heather does not occur (quadrats 11-18).

> **A good point. The student could have suggested some ways in which these plants may be competing with each other**

Possible skills assessed in this coursework

'Handling apparatus and materials' (Ch. 2, Unit 3), 'Observing and measuring' (Ch. 2, Unit 4), 'Recording and communciation' (Ch. 2, Unit 5), 'Intepreting information' (Ch. 2, Unit 6).

Examiner assessment

High standard. The information is presented in a well-organised way. The student made good use of the results, and has demonstrated a very good understanding of the theoretical basis of the feeding relationships of the organisms in the sample. The criticisms of the experiment are valid and fairly thorough.

Suggestions for further work

This exercise: Include vertebrate animals and also producers (i.e. green plants) in the sample. Examine and draw selected specimens to show adaptations to their feeding behaviour, e.g. mouthparts. Find out the sensitivity of different species to oxygen concentration, water flow, temperature, pH, etc.

Related exercises: Compare the species diversity and abundance in two different freshwater ecosystems; differences could include slow-/fast-moving, polluted/unpolluted, deep/shallow, etc. Use a capture-mark-release-recapture system for estimating the size of a population, e.g. of freshwater shrimps.

Linked work: Natural classification (Unit 5), Artificial classification (Unit 6), Variation (Unit 12), Climatic factors (Unit 18).